CALIFORNIA SPRING WILDFLOWERS

California
Spring
Wildflowers

FROM THE BASE OF THE SIERRA NEVADA
AND SOUTHERN MOUNTAINS TO THE SEA

by Philip A. Munz

.

UNIVERSITY OF CALIFORNIA PRESS
BERKELEY, LOS ANGELES, LONDON

UNIVERSITY OF CALIFORNIA PRESS
Berkeley and Los Angeles, California

UNIVERSITY OF CALIFORNIA PRESS, LTD.
London, England

ISBN: 0-520-00896-0
Library of Congress Catalog Number: 61:7524
Printed in the United States of America

9 0

CONTENTS

INTRODUCTION

California has long been considered an earthly paradise, especially in the spring when its rolling hills and green valleys are full of wildflowers. There are perhaps 6,000 flowering plants in the state, many of which, like the grasses and sedges, are very important for grazing but not of especial interest to the wildflower lover. However, even when these and the trees and the more inconspicuous bushes are deleted from the list, some thousands of real wildflowers still remain. The various gilias or penstemons or paint-brushes, among many, are so much alike that only the more observant and perhaps technically interested person is going to want to differentiate them. Therefore, in a wildflower book these can be treated in groups. The more discriminating student can turn to A CALIFORNIA FLORA by Munz and Keck (University of California Press, 1959) for more detail.

When we recall the great variety of topographical conditions in California and the plants we see in its different areas, we know that the desert flowers are quite different from those on coastal slopes and that the summer bloomers in high mountains differ from the spring plants of the valleys. Therefore, to bring before the public in compact and useful form something by which wildflowers can be identified, this little book is presented. It includes only spring bloomers and those from the foothills of the Sierra Nevada and more southern mountains and between these foothills and the coast.

CLIMATIC CONDITIONS

Between the mountains and the coast the topography exhibits considerable range. Some of it is wooded, some brushy, some is grassland. But all of it agrees in the general climatic pattern that has come down for some period of time geologically, producing a vegetation quite characteristic and often spoken of as a Mediterranean type. The moisture comes overwhelmingly in the cooler winter months and is followed by a long dry period that is very hot toward the interior and cooler only near the coast, where the fogs and humidity of the ocean air help to prolong the growth season much more than in the hot interior. In either instance, at lower altitudes, snow falls in small amounts or not at all in winter and the flowering season is in spring, with little or no bloom in summer except along streams or about seeps and ditches. In the Yellow Pine Belt and above, there is winter snow and the seasons are more like those in our more eastern and northern states.

This book deals with the area below the Yellow Pine and between it and the coast. It is an area of variable precipitation, from about ten inches in the neighborhood of San Diego and parts of the Central Valley to about one hundred in the extreme northern Coast Ranges. Usually, grassland prevails where the rainfall is between six and twenty inches, shrubby growth or chaparral or scrub between fifteen and twenty-five inches, woodland between twenty and forty inches, and a denser forest occurs with higher rainfall, especially nearer the coast where the air is cool. These plant formations are not sharply separated by precipitation, but often are by topography. Gently rolling hills may have grassland and, with a little more moisture, open woodland, while on nearby stonier and steeper slopes may appear chaparral or other brush.

Our broad-leaved evergreen trees and shrubs such as oaks and California-lilacs tend to have very harsh leaves with rather reduced surfaces as compared with their relatives in regions with summer rains, thus cutting down evaporation. Others may lose their leaves in the dry season, as does the California Buckeye (see page 34). Still others, like the Canyon Maple (see page 99), grow only where their roots have access to moisture at all seasons. Over all, our California conditions produce much open country that becomes green with the advent of the rains in late fall or early winter. Seedlings of flowering annuals develop slowly through the winter as does the new growth on shrubs and trees. The great season of flowering is from February to April or even May. Then brownness and dormancy again set in and the summer is largely a period of inactivity.

How to Identify a Wildflower

For identification it is most helpful to have flowers available and not just the vegetative parts of the plant. To refresh the reader's memory there are given herewith the parts of a typical flower (fig. A) showing: (1) the outer usually greenish *sepals*, known collectively as the *calyx;* (2) the inner usually colored *petals*, which taken together constitute the *corolla;* (3) the *stamens*, each typically with an elongate basal portion, the *filament*, and a terminal more sacklike part in which the pollen is produced, the *anther;* and (4) the central *pistil*, with a basal enlarged *ovary* containing the immature seeds, an elongate *style*, and one or more terminal *stigmas*, on which the pollen grains fall or rub off an insect or humming bird. These many parts may be greatly modified. The sepals may be separate, more or less united, and alike or not alike. The same is true of the petals. The corolla may consist of separate similar petals. Petals may be reduced or quite lacking, or they may be united to form tubular, often two-lipped, structures that afford landing platforms for bees and other visitors, in which case the stamens and style may be arched over so as easily to deposit pollen on or receive it from the body of the insect. The ovary may be partly sunken into tissues below or fused with them in such fashion as to be evident below the flower instead of up in it. When a person looks at a flower, he should observe such conformations, and should pay some attention to the number of parts of a given series, as the petals. Superficially, the blue flower of a *Gilia* may resemble that of a *Brodiaea*, but the former will show five petal parts, the latter has six segments. In other words, it is necessary at times to examine flowers in detail and with care.

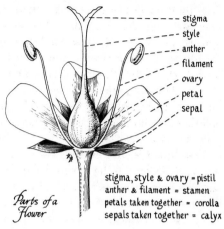

stigma, style & ovary = pistil
anther & filament = stamen
Parts of a Flower
petals taken together = corolla
sepals taken together = calyx

FIGURE A.   A REPRESENTATIVE FLOWER

The wildflowers illustrated by line drawings in the text are arranged in four sections: (1) whitish to pale cream or with a pink flush; (2) red to rose or reddish purple or reddish brown; (3) blue or lavender; and (4) yellow to orange or greenish yellow. Identification of a red flower, for example, can be attempted by leafing through the drawings of the red-flowered section and by use of the index to the color-plates, and can be further aided by reading the text opposite the page of illustration. Such an artificial arrangement is helpful, but will not always work. A lupine flower that is bluish when young and fresh often changes to reddish as it grows old and is about to die. Then, too, individuals of the same species can vary greatly in color, say from blue to lavender or almost white. Almost every species has albino forms — white variants for red and blue types, yellow for those normally red or scarlet. Remember that you must allow for variation in color, for change with age, and for the fact that in nature growth is not fixed but fairly inconstant.

Perhaps a word should be added as to the richness of the California flora. The long dry season when dormancy is the rule is accompanied in the wet part of the year by a great wealth of annuals, many of which are highly colored. Since David Douglas came to the state more than a century ago to seek bulbs and seeds for introduction into England, our plants have been much prized in northern Europe. They have done well there because of the cool nights like our own and have been much used. Godetias, baby–blue–eyes, and gilias, for instance, commonly met in gardens there, have often been greatly developed horticulturally, with many color forms, double flowers, and the like. But with the increase in population in California and with the encroachment on wild lands by industry and agriculture, many California wildflowers are increasingly rare. They need protection if we and our children are to enjoy them. Turning the camera on them rather than picking them is the means to permanent enjoyment.

Many of the plants that now seem to us a natural constituent of the California landscape were not here when the first Europeans arrived. I refer to the mustards, filarees, wild oats, and many others, some of which may be quite weedy and unattractive (Tumbleweed, Russian-Thistle, Purslane) or may be showy and add color to our fields and orchards (Mustard, Oxalis, Foxglove).

In this book nomenclature follows that of A CALIFORNIA FLORA by Munz and Keck, to facilitate cross reference. Occasional notes on early use by former inhabitants of the state or on folklore are given. Mention, too, of species other than those pictured should help identify more than the number actually illustrated here.

ACKNOWLEDGMENTS

Most of the drawings used in this book are by Dr. Stephen S. Tillett, now of New York Botanical Garden; a smaller number are by Professor Richard J. Shaw of Utah State University. The majority of the Kodachromes used are the property of the Rancho Santa Ana Botanic Garden and are the work of several persons, chiefly Percy C. Everett and Richard J. Shaw. The following were kindly loaned by Miss Beatrice F. Howitt of Pacific Grove, California: Yellow Sand-Verbena, Meadow-Foam, Prickly Poppy, Godetia in Plate 48, *Rhododendron macrophyllum, Collomia grandiflora, Epipactis gigantea,* and *Trillium chloropetalum* and *ovatum.*

To the foregoing people and to Gladys Boggess, secretary at the Botanic Garden, I express with great pleasure my gratitude for their help in securing materials and in preparation of the manuscript.

FLOWERS WHITE TO PALE CREAM
OR PALE PINK OR GREENISH-WHITE

Section One

About marshes and wet places, particularly in central and northern California, there is found a group of plants with white flowers and sheathing leaf-bases. Their flower stalks project up out of the water and bear branching clusters with flowers about the size of a quarter or a half dollar. The leaves are long-stalked and are expanded into more or less egg-shaped or arrowlike blades one to five inches long. In the FRINGED or STAR WATER-PLANTAIN (*Machaerocarpus californicus*), figure 1, the leaves are not arrowlike and the seed-bearing structures spread out like a star, having long beaks at their tips. In the common Water-Plantain (*Alisma Plantago-aquatica*), the leaves are the same but the seed-bearing structure has a little short beak to one side. In the Arrowhead (*Sagittaria*) the leaf is usually arrow-shaped and the upper part of the flower cluster bears only male flowers with stamens and the lower part only female flowers that produce seeds.

FIGURE 1. STAR WATER-PLANTAIN

Another group of marsh plants, which might be confused with these and look as if they were related, includes BUR-REED (*Sparganium*), figure 2. They have much the same habit, but the leaves are long and narrow; there are no petals. The male flowers in the upper part of the flower cluster soon dry up and the lower fertile flowers produce rounded clusters of many burs, each of which contains one seed.

FIGURE 2. BUR-REED

A member of the Lily Family with small flowers in rather large branching terminal clusters is SOAP PLANT or AMOLE (*Chlorogalum pomeridianum*), figure 3. Its rather large underground bulb is covered with dark brown fibers of the old bulb coats. It contains a saponin and early settlers used it commonly for shampoos, while the Indians roasted and ate its abundant starch. The long narrow leaves appear after the first

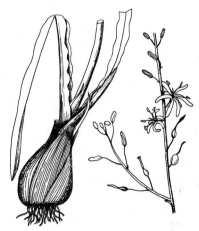

FIGURE 3. SOAP PLANT

FIGURE 4. FALSE SOLOMON'S-SEAL

FIGURE 5. ZYGADENE

FIGURE 6. ADDER'S-TONGUE

autumn rains and remain for many months, a conspicuous basal cluster with wavy margins. The stems branch freely above, grow to be five to eight feet tall, and bear the many small white lilylike flowers. It is found in dry open places below 5,000 feet, from San Diego to southern Oregon.

Common from the Pacific Coast to the Atlantic are various forms of FALSE SOLOMON'S-SEAL; these are perennial herbs with thick creeping underground rootstocks. The ascending stems produce lance-shaped to egg-shaped leaves in two rows and a terminal cluster of small white flowers followed by reddish or red-purple berries. Our most common kinds in California are *Smilacina racemosa* var. *amplexicaulis* with many flowers in a branched cluster and S. *stellata* var. *sessilifolia*, figure 4, with fewer flowers in a simple cluster. These or other forms occur in much of our area, especially in partly shaded places below 6,000 feet.

ZYGADENE or STAR-LILY (*Zigadenus Fremontii*), figure 5, is another bulbous plant with narrow basal leaves and a flower stalk branched above. It has small yellowish-white flowers bearing greenish-yellow glands near their center. It is found on grassy or bushy slopes, often most conspicuously after a fire, from San Diego County to Oregon. Other species are poisonous to cattle.

In California are recognized twelve species of ADDER'S-TONGUE or FAWN-LILY, which vary from white to yellow or rose. They have deep-seated solid bulblike structures with membranous coats and two leaves near the base of the stem. The leaves may or may not be mottled. The species shown here is *Erythronium multiscapoideum*, figure 6, with mottled leaves and whitish flowers. It grows in woods from Mariposa County to Tehama and Butte counties and blooms from March to May.

One of the most thrilling groups of California wildflowers is that of the Mariposa-Lily or the genus *Calochortus*, of which we have about thirty-five species. These are of several types: one with small closed pendent flowers, another with mostly larger, more open, and bowl-shaped ones. Here, in figure 7, is an example of the former sort, the FAIRY LANTERN or GLOBE-LILY (*C. albus*). One of the daintiest of wildflowers, with its nodding cluster, it rises from a basal group of two to six long leaves. It grows in shaded, often rocky, places in woods and canyons below 5,000 feet, from San Diego County to San Francisco and in the foothills of the Sierra Nevada. In the Santa Lucia Mountains and other places it takes on a rose tinge and looks quite different. For the more usual and bowl-shaped type see page 81.

The modern botanist recognizes another family near the Lily Family, namely the Amaryllis Family. Its flowers are much the same, with three outer and three inner petallike parts, but the flowers are arranged in an umbel, that is, they all grow out from one point instead of being up and down a central axis. Here belong daffodils and Chinese-lilies and, among the wild plants, brodiaeas and onions (*Allium*). A related plant is MUILLA (anagram of *Allium*), but it lacks the onion taste and odor. The common species is *Muilla maritima*, figure 8, found on flats and slopes from Glenn County south in the Coast Ranges and to some extent in the Sacramento and San Joaquin valleys.

In low wet usually alkaline places grows a peculiar plant with large leaves and long runners that give rise to new individuals. The stems are about four to twenty inches high and end in what most people assume to be a flower with a number of white

FIGURE 7. FAIRY LANTERN

FIGURE 8. MUILLA

FIGURE 9. YERBA MANSA

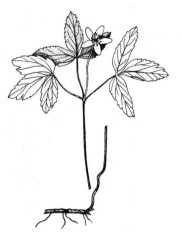

FIGURE 10. ANEMONE

FIGURE 11. CREAM CUPS

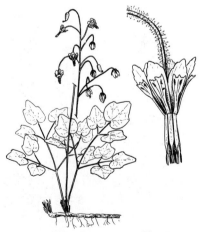

FIGURE 12. INSIDE-OUT FLOWER

petals near the base. Close examination will show that these are modified leaves and that there are smaller similar ones all the way up the spike, which is really made up of many small flowers. This plant is called by the Mexicans YERBA MANSA (*Anemopsis californica*), figure 9, meaning soft herb. Its range extends from Sacramento Valley and Santa Clara County to Lower California and Texas.

Superficially the foregoing *Yerba Mansa* suggests an Anemone, as indicated by its scientific name, but the true ANEMONE has compound or lobed leaves and the flower is a single true one with petallike structures around one set of stamens and pistils. The species pictured here, *Anemone quinquefolia* var. *Grayi*, figure 10, is a plant of wooded slopes in the Redwood region from Santa Cruz County to Sonoma County.

California has many kinds of poppies, a group that adds much color to the fields and slopes of most of the state. One of the loveliest of our spring flowers is CREAM CUPS (*Platystemon californicus*), figure 11, a soft-hairy little annual with flowers from almost white to cream or yellow, or it may have a reddish tinge. It is found in open sandy or clayey areas, on burns and disturbed places in most of our state.

INSIDE-OUT FLOWER (*Vancouveria planipetala*), figure 12, bears the name of Captain George Vancouver who visited the California coast in 1792. It is related to the barberries and Oregon-Grape, but is a perennial herb with a running rootstock and bears compound leaves and small yellow or, in the species shown here, white flowers. It grows in the shade of woods from Monterey County north.

Related to the poppies but with much simpler flowers are the mustards. They have four petals, a biting peppery sap, and many, such as the Radish, Cabbage, and

Cauliflower, are important food plants.
California has many native plants in this
family, some of which have white flowers
and are shown here. LACE POD or FRINGE
POD (*Thysanocarpus*) is a small group of
annuals with slender erect stems and
rounded almost flat pods with a scalloped
or even perforated wing, which may be
colored and quite handsome.

The one illustrated in figure 13, *T. curv-
ipes,* has clasping stem-leaves and is common in grassy or brushy places from Lower
California to British Columbia. *Thysanocarpus laciniatus* looks much the same, but
is more delicate and the stem-leaves are not
clasping at their base.

SWEET ALYSSUM (*Lobularia maritima*),
figure 14, is a common garden plant and
was introduced at an early date from Europe
for its fragrance. It has escaped and established itself in much of California. Another
waif that is found in almost every field and
vacant lot, and was originally brought over
from Europe, is SHEPHERD'S PURSE (*Cap-
sella Bursa-pastoris*), figure 15. It has little
of beauty, but is so common that it will be
seen by everyone.

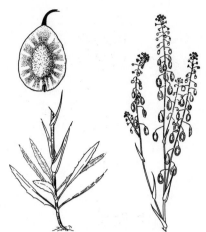

FIGURE 13. LACE POD

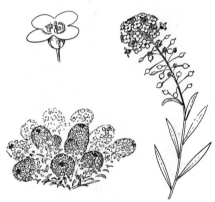

FIGURE 14. SWEET ALYSSUM

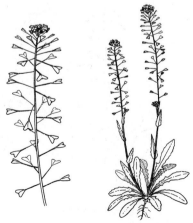

FIGURE 15. SHEPHERD'S PURSE

FIGURE 16. THELYPODIUM

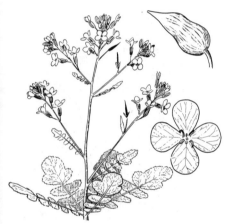

FIGURE 17. WILD RADISH

FIGURE 18. TOOTHWORT

Quite a different lot in the Mustard Family are the long-podded plants like Mustard itself. One of the natives is the THELYPODIUM group of rather weedy plants mostly with whitish flowers. They are largely erect annuals and are widely distributed. An example is *T. flavescens*, figure 16, of the inner Coast Ranges from San Benito County to Solano County.

The common WILD RADISH (*Raphanus sativus*), figure 17, is another weed in waste places and fields; the flowers are commonly white with reddish or purplish veins. It is naturalized from Europe and the Garden Radish is a cultivated form. TOOTHWORT or MILK MAIDS (*Dentaria californica*), figure 18, is a native and is found on shaded banks and slopes from San Diego County to Oregon. It is variable and some forms have simple, some have compound, leaves; the petals are about one-half inch long. Flowering commonly begins in February; it is, in fact, one of the first perennials to bloom. The roots bear small tubers.

The Purslane Family, to which belong our Garden Portulaca and the weed Purslane, is quite a family in California. It is characterized by rarely having more than two sepals and by being quite fleshy. One of the earliest flowers in spring is BITTER-ROOT (*Lewisia rediviva*), figure 19, with a thick fleshy taproot from the crown of which appears a tuft of white flowers that may age quite pink. Later come the fleshy leaves. Here it grows in rocky and gravelly places from the San Bernardino Mountains northward. It is the state flower of Montana.

FIGURE 19. BITTERROOT

Every Californian is familiar with Miner's-Lettuce, a green annual with a pair of stem-leaves grown together to form a cup, and with small white flowers. It belongs to the genus MONTIA, of which there is shown in figure 20, *M. sibirica*, a rather fleshy perennial found from Fresno and Santa Cruz counties northward and not having the pair of stem-leaves grown together.

The Pink Family has many showy plants, for example the garden pinks and carnations. Others like the chickweeds are less showy, but very common. One, a perennial CHICKWEED (*Stellaria littoralis*), figure 21, is found in moist places in coastal sands from San Francisco to Humboldt County. Its small flowers have divided petals.

FIGURE 20. MONTIA

FIGURE 21. CHICKWEED

FIGURE 22. SANDWORT

FIGURE 23. SAND-SPURREY

FIGURE 24. LIVE-FOREVER

SANDWORT (*Arenaria macrophylla*), figure 22, is related to the preceding Chickweed. It is a perennial, growing about eight inches high and found on shaded slopes in the northern Sierra, in the Coast Ranges from Mount Hamilton north, and in the Cuyamaca Mountains of San Diego County. Other closely related species in the state may be annual, may have narrower leaves, and short or long petals.

Another group, also with the characteristic paired leaves of the family, is that of the sand-spurreys. They are often found in somewhat salty places near the beach. One of the most prominent is the common SAND-SPURREY (*Spergularia macrotheca*), figure 23, growing near coastal salt marshes the length of the state. The flower is about half an inch across; at the base of the leaf are the somewhat papery translucent stipules and the main leaves have bundles of smaller ones in their axils.

Many persons are fond of succulents, of which there are various groups. One is the Stonecrop Family to which belong Live-Forever, Hen-and-Chickens, and similar plants. Many of our California plants belong to the genus *Dudleya*, named for Professor W. R. Dudley of the original faculty at Stanford University. Here is shown *Dudleya virens*, figure 24, a LIVE-FOREVER coming mostly from the Channel Islands, but distinguishable only with difficulty from several species on the mainland. Its leaves are just slightly flattened, but other species have them much more so. Some have green leaves, some gray, and some almost white with a thick powder.

The Saxifrage Family with us is largely montane and in midsummer many species occur in the high Sierra, but there are some spring bloomers at lower elevations also. One of these is the common SAXIFRAGE (*Saxifraga californica*), figure 25, found on shaded often grassy banks from Orange County north through the Coast Ranges and Sierran foothills to Oregon. Growing to about a foot tall, it is usually much smaller and the flowers are scarcely the size of a dime.

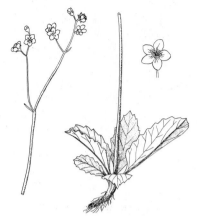

FIGURE 25. SAXIFRAGE

A plant of much more heroic stature is the UMBRELLA PLANT (*Peltiphyllum peltatum*), figure 26, found along the banks of rapid streams, which attains the height of about three feet. It grows along the western base of the Sierra Nevada from Tulare County north and in the Coast Ranges of Humboldt and Siskiyou counties. The young growth was eaten by the Indians.

The WOODLAND STAR is represented by several species, one of the most common, *Lithophragma heterophylla*, figure 27, occurring on shaded slopes below 6,500 feet, from San Diego County to Oregon. The petals may be entire to deeply toothed.

FIGURE 26. UMBRELLA PLANT

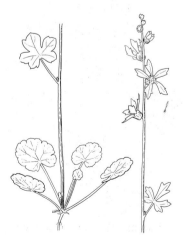

FIGURE 27. WOODLAND STAR

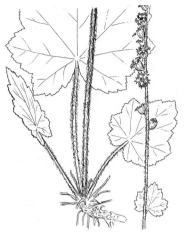

FIGURE 28. FRINGE CUPS

FIGURE 29. HONEY DEW

FIGURE 30. CALIFORNIA BLACKBERRY

Another member of the preceding Saxifrage Family is FRINGE CUPS (*Tellima grandiflora*), figure 28, whose petals are divided into very narrow segments. The plant reaches to about thirty inches in height and is quite hairy. It is found in moist woods and rocky places, below 5,000 feet, from San Luis Obispo and Placer counties north to Alaska.

In the Rose Family are plants of many kinds, from small herbs like the Strawberry to trees like the Apple and Pear. One of our white-flowered members is HONEY DEW (*Horkelia cuneata*), figure 29, a glandular herb with flat open flowers about one-half inch across. Between every two sepals is a bract or reduced leaf, so that it looks as if there were ten instead of five sepals. The stamens are many. The plant is found in open sandy places and in woods, from San Francisco to San Diego.

Growing in somewhat similar but usually damper places is a Bramble or CALIFORNIA BLACKBERRY (*Rubus vitifolius*), figure 30, with flowers up to an inch across. The berries are usually disappointing, since our dry summers seem not to promote the development of very juicy fruit. It grows near the coast from Mendocino to San Luis Obispo counties and is difficult to separate from *Rubus ursinus*, which ranges from Oregon to Lower California and from which the Loganberry, Youngberry, and Boysenberry have been developed. *Rubus ursinus* has duller leaves that are more woolly beneath than has *R. vitifolius*.

Among the woody members of the Rose Family are the cherries, plums, and peaches, all of which have stone fruits, that is, a large central hard seed covered with a fleshy layer. These belong to the genus *Prunus* and in California we have among others the HOLLY-LEAVED CHERRY (*P. ilicifolia*), figure 31. It is a handsome shrub with shining stiff leaves having wavy spinose-toothed margins. The flowers are white and in short spikes; the fruits red, about one-half inch in diameter and with thin sweetish pulp. It is common on dry slopes and fans below 5,000 feet from Napa County to Lower California.

FIGURE 31. HOLLY-LEAVED CHERRY

A more deciduous shrub is the SERVICE BERRY, also with several species, the most common being *Amelanchier pallida*, figure 32. The flowers are almost an inch across, with strap-shaped petals; the fruit is round, purplish-black. The species makes bushes three to eighteen feet high and is found scattered through the Coast Ranges and Sierra Nevada from San Diego County to Oregon.

FIGURE 32. SERVICE BERRY

We have in California over eighty species of lupines. They grow everywhere from the coastal strand to the highest mountain tops, from the sea to the deserts. One of the most common of the white species is LUPINUS DENSIFLORUS, figure 33, an annual, rather fleshy, about one foot tall, with flowers in definite whorls. It has yellowish and pinkish forms and other variations. It grows in grassy and open places, often on new road cuts, from Humboldt and Butte counties to western San Diego County.

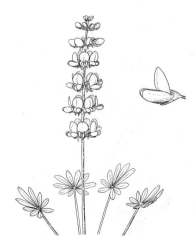

FIGURE 33. LUPINE

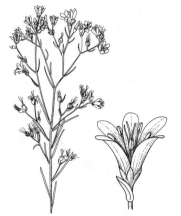

FIGURE 34.   FLAX

FIGURE 35.   BUCKBRUSH

FIGURE 36.   FLOWERING DOGWOOD

Flax to most of us is a fairly large blue-flowered plant of the pine belt and with stems so exceedingly tough that when we try to pick only a flower, we usually come away with the whole plant. These fibers in the outer stem are, of course, the source of linen in an Old World species. In our Coast Ranges we have a number of small-flowered white species of FLAX, among them *Linum californicum*, figure 34, which grows in grassy and rocky places from Glenn and Butte counties to San Benito County.

Among the California shrubs no group is more diverse and fascinating than the California-Lilac or Buckbrush or Ceanothus. These so-called lilacs are usually blue, while the buckbrushes are mostly white and constitute a distinct section of the genus *Ceanothus*, having corky bumps on the twigs at the base of each leaf. Representative of this BUCKBRUSH group is *Ceanothus megacarpus*, figure 35, of chaparral from Santa Barbara to San Diego. Closely related species vary in leaf-shape, whether leaves are notched at the tip or not, whether the seed-capsule is horned or not, and how large it becomes. These species are found on dry slopes and fans in much of the state. See also page 41.

FLOWERING DOGWOOD on the West Coast is a much less graceful bush than in the Eastern states; nonetheless it is a pleasant plant to find in mountain woods. It may grow to tree-size farther north; in general it is found in brush and woods below 6,000 feet from San Diego County to British Columbia. The large petallike structures are whitish leaves surrounding a whole cluster of very small flowers; the fruits are red, about a half-inch long. The species illustrated in figure 36 is *Cornus Nuttallii*.

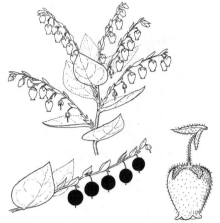

In the famous Tournament of Roses Parade held on New Year's Day in Pasadena, many of the floats have a background covering of leathery green leaves known in the trade as "lemon leaves." These come from a low spreading subshrub of our north coast, the SALAL (*Gaultheria Shallon*), figure 37, related to the Wintergreen of Eastern woods. Our plant makes a veritable ground cover in forests in northern California and grows near the coast as far south as Santa Barbara County. Its dark purple fruit is striking.

FIGURE 37. SALAL

Related to it and with the same urn-shaped flowers is Manzanita, one of California's most interesting groups with more than forty species in the state (see pages 29 and 76). Varying from prostrate plants like the Bearberry of the north coast to erect and almost treelike shrubs in the interior, some are hairy, some smooth, some exhibit leaflike bracts among the flowers, others have much reduced ones. All have the same petite pink or white flowers and a reddish fruit resembling a small apple, which is what the Spanish word MANZANITA means. The species here illustrated, *Arctostaphylos insularis*, figure 38, resembles most others in its clear reddish-brown bark and stiff leaves.

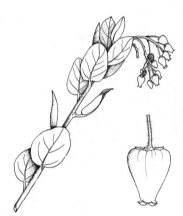

FIGURE 38. MANZANITA

There are in California two rhododendrons; the one shown here is WESTERN AZALEA (*Rhododendron occidentale*), figure 39, a deciduous shrub with large white flowers usually having a tinge of yellow. It is found on stream banks and in moist places in the mountains of San Diego and Riverside counties and at lower elevations in the Coast Ranges and Sierra Nevada from Santa Cruz and Kern counties north. It is said to be poisonous to stock. For the other species see page 85.

FIGURE 39. WESTERN AZALEA

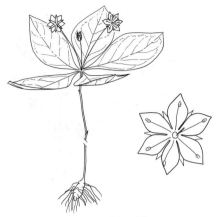

FIGURE 40. STAR FLOWER

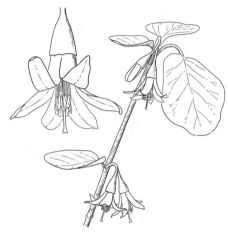

FIGURE 41. STORAX

FIGURE 42. MORNING-GLORY

STAR FLOWER (*Trientalis latifolia*), figure 40, is a small perennial with its five to seven petals grown together at the base. The plant comes from a tuberous rootstock and sends up an erect stem with a whorl of leaves. In California it is found in shaded woods from San Luis Obispo and Mariposa counties north.

In canyons and rocky places among chaparral shrubs or in woods there grows a sweet-smelling bush with pendulous white flowers that appear in April and May. This is STORAX or SNOWDROP BUSH (*Styrax officinalis* var. *californica*), figure 41. The flowers appear with the young leaves. More than one form occurs in the state, but the various forms are only technically distinguishable. They occur from San Diego County to Shasta and Lake counties. The gum storax comes from a Mediterranean species.

The Common Bindweed, introduced from Europe, has become a great pest in orchards and fields and has turned us against plants with morning-glory-like flowers, but our native species are quite harmless and interesting. With funnel-shaped blossoms, white with a tinge of pink, these plants occur widely, clambering over bushes or growing flat on the ground. The most common MORNING-GLORY (*Convolvulus occidentalis*), figure 42, is found throughout our area and bears flowers more than an inch long.

One of the world's centers for the Phlox
Family is California and some of the forms
found here are attractive plants, most of
them being pink or rose but some almost or
quite white. Among the latter is a group
largely of annuals with opposite or paired
leaves divided to their base into narrow
linear segments. They are LINANTHUS, one
of which, *L. grandiflorus*, figure 43, becomes a foot or so tall and has white to
pale lilac flowers one-half to one inch long.
It is found in open woods and sandy places
below 3,500 feet from the shore into the
Coast Ranges and from Sonoma County to
Santa Barbara County.

FIGURE 43.   LINANTHUS

Another type of LINANTHUS is a much
more open type, with the flowers not in
dense clusters. The one shown in figure 44
is the *Linanthus liniflorus* that grows to
almost two feet in height and is found in
open dry places below 5,000 feet from
southern California to Washington. The
flowers extend to about an inch in length.
For another species of *Linanthus* see page
36.

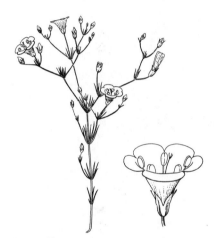

FIGURE 44.   LINANTHUS

In figure 45 is a member of the same
family, but of the genus NAVARRETIA.
These are more or less spiny plants and
many occur in the dry mud of depressions
that have held water during the rainy
season, that is, vernal pools. The plant
drawn, *Navarretia leucocephala*, is found
in the foothills of the Sierra Nevada from
Eldorado County north and in the Coast
Ranges from San Benito County north.
This and other allied species may not develop a central stem, but form flat masses
on the ground.

FIGURE 45.   NAVARRETIA

FIGURE 46. WILD-HELIOTROPE

FIGURE 47. HELIOTROPIUM

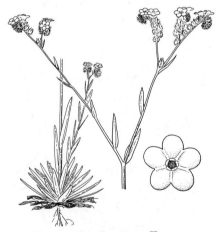

FIGURE 48. POPCORN FLOWER

In the California spring flora are many annuals in which the flowers are arranged in clusters that coil downward at the tip somewhat like the coil of a snail shell. Among these is the so-called WILD-HELIOTROPE, genus *Phacelia,* of which we have a number of white-flowered species, like *P. Douglasii,* figure 46. It is an annual, mostly with spreading or prostrate stems and with flowers about half an inch wide. It is found in sandy places from Riverside County to central California. Other usually quite hairy or even bristly species, annual or perennial, often erect, with lobed or almost entire leaves, are in most parts of the state.

With the same coiled flower clusters is a related family to which the genus HELIO-TROPIUM itself belongs, example *H. curas-savicum,* figure 47, whose flowers become rather purplish in the center. It is perennial, fleshy, and occurs in somewhat saline places throughout California.

In the same family is the POPCORN FLOWER (*Plagiobothrys nothofulvus*), figure 48, of open grassy places. The lower part of the plant has a purplish dye and the hairs about the flowers are quite tawny.

Related to Popcorn Flower, but more stiff-hairy, without purple dye, and mostly with smaller flowers is a series of annuals sometimes called popcorn flowers, but perhaps better WHITE FORGET-ME-NOT (*Cryptantha intermedia* and other species), figure 49. They are common in the spring in dry open places, like burns in the chaparrel, through much of the state.

Apparently without a common name, but commonly enough found and often in great masses, is a group of small-flowered plants, mostly prostrate and with conspicuous spreading fruits made up of four spreading nutlets, the genus PECTOCARYA, figure 50. They occur mostly in sandy or gravelly places and flower from March to May.

The Mint Family is one noted for its many fragrant members because of their essential oils, such as Mint, Bergamont, Thyme, Pennyroyal, Catnip, and so forth. Among our natives are several interesting species of Sage, such as WHITE SAGE (*Salvia apiana*), figure 51, an important bee plant, as indicated by its scientific name. Shrubby below and reaching a height of several feet, it is characterized by white leaves and openly branched flower clusters. It grows from Santa Barbara County to Lower California.

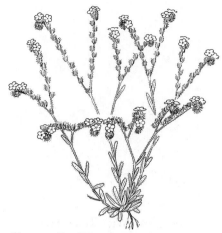

FIGURE 49.   WHITE FORGET-ME-NOT

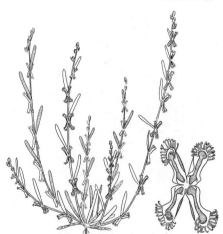

FIGURE 50.   PECTOCARYA

FIGURE 51.   WHITE SAGE

FIGURE 52   YERBA BUENA

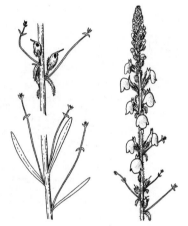

FIGURE 53.   SNAPDRAGON

FIGURE 54.   WILD-CUCUMBER

A plant having a fragrance more delicate than the White Sage of figure 51 and growing flat on the ground is the little YERBA BUENA (*Satureja Douglasii*), figure 52. With slender trailing stems and leaves up to about an inch long, it bears small white or purplish flowers. The odor of the crushed foliage is very pleasant. It has been used by the Indians and early settlers as a tea for drinking and as a medicine for fevers and stomach complaints.

The Snapdragon Family has many beautiful plants and like the mints is characterized by a two-lipped flower. One of the common white members is a SNAPDRAGON (*Antirrhinum Coulterianum*), figure 53, growing to about four feet high, in open often disturbed places like burns, from Santa Barbara County south.

California has not many native vines, but among them is the interesting WILD-CUCUMBER or BIG-ROOT or MAN ROOT (*Marah fabaceus*), figure 54. The part above ground dies back each summer and may make a great mass fifteen to twenty feet high, clambering over bushes and almost smothering them. The root structure is remarkable, lasting from year to year, shaped like a huge turnip or beet, and weighing many pounds. It stores much water and starch. The female flower produces a spiny fruit a couple of inches long with large brownish seeds. This species or related ones can be found among bushes in much of California.

The Honeysuckle Family is characterized by having its leaves in pairs (opposite)
and often by two-lipped flowers. But in
California conspicuous members of the
family often have the flowers regular, that
is, the halves are alike when bisected by a
vertical plane, no matter in which direction
it is passed. Among these latter are shrubs
such as Snowberry (*Symphoricarpos
mollis*), figure 55, low partly trailing
bushes with flowers about one-fourth inch
long and with rather spongy white fruits
almost twice as long. Common on shaded
slopes, it is found mostly below 3,000 feet
in the Coast Ranges from Mendocino
County to Lower California.

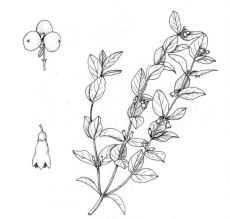

FIGURE 55.   SNOWBERRY

Another shrub is Elderberry (*Sam-
bucus mexicana*), figure 56, with compound leaves and large flat-topped clusters
of small white flowers. It gets to be a large
bush or small tree, almost deciduous in the
dry season especially in the southern part
of its range. The berries are blue or white,
with a white powdery coating. It is found
from Lake and Glenn counties southward.
Other species of Elderberry occur in the
mountains and to the north.

FIGURE 56.   ELDERBERRY

In figure 57 is shown the Western
Coltsfoot (*Petasites palmatus*), of the
Sunflower Family. It is a perennial herb
with creeping rootstocks and sends up in
early spring a stem with large leaflike
bracts and a terminal dense cluster of
heads. When these are examined closely
they are found to consist of many minute
flowers surrounded by a cuplike involucre
of many bracts. The large lobed leaves
appear later. Coltsfoot is found in deep
shade in the Coast Ranges from the Santa
Lucia Mountains northward, even to
Alaska.

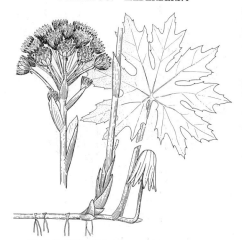

FIGURE 57.   WESTERN COLTSFOOT

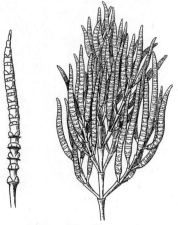

In coastal salt marshes and low alkaline places from the San Joaquin Valley and from San Francisco Bay to Mexico is a relative of Saltbush (see pages 51 and 96), namely SAMPHIRE or GLASSWORT (*Salicornia subterminalis*), figure 58. It grows to a height of about one foot, has leafless fleshy jointed stems and terminal spikes with reduced sunken greenish flowers.

FIGURE 58. GLASSWORT

PLATE 1. OCEAN SPRAY

In woods and rocky places near the coast is a slender-stemmed shrub with masses of small spiraealike flowers that range from whitish to pale pink. It is CREAM BUSH or OCEAN SPRAY (*Holodiscus discolor*), plate 1, and comes in a number of forms that vary in size of leaves, teeth on the leaves, and the like. It extends from Orange County north to British Columbia. Flowers appear from April to July.

PLATE 2. BEAVER TAIL

BEAVER TAIL (*Opuntia basilaris*) is shown in plate 2. It is a so-called spineless cactus, but does have many short stiff prickly hairs, glochids, that are very difficult to remove from a person's skin. It has different growth forms, but is mostly a low cactus, often with very handsome somewhat purplish joints. The form west of the deserts is variety *ramosa* and is found on slopes and washes at scattered stations from the Greenhorn Mountains in Kern County to the Vulcan Mountains in San Diego County. It blooms mostly in May.

PLATE 3. MEADOW-FOAM

One of the most distinctively Californian wildflowers is MEADOW-FOAM, plate 3 (*Limnanthes Douglasii*). An annual, with dissected leaves and yellow petals with white tips, it forms plants often a foot or so across. It grows in great masses in moist places, particularly in the inner Coast Ranges from San Luis Obispo County to southern Oregon and blooms from March to May. Yellowish forms appear nearer the coast and pinkish ones in the Sierran foothills. Related species are found as far south as the Laguna Mountains of San Diego County.

27

PLATE 4. WILD SWEET PEA

PLATE 5. SAND-VERBENA

PLATE 6. PRICKLY-PHLOX

We have in California twenty-two species of SWEET PEA or *Lathyrus* reported as growing wild; some have escaped from gardens, others are native. One of the latter is shown in plate 4 (*Lathyrus vestitus* subspecies *puberulus*), which is found climbing over bushes in the chaparral from Humboldt to Los Angeles counties and blooming from April to June. Other species of Wild Sweet Pea have white to yellowish flowers, some a deep rich red, but our native perennials do not have the pleasant odor of the cultivated Sweet Pea.

SAND-VERBENA, plate 5, is mostly a large perennial plant forming mats several feet across. In our area three species are most common and they are found along the beach and dunes back of it. The yellow-flowered one here shown (*Abronia latifolia*) ranges from Santa Barbara County to British Columbia and blooms from May to October. Its leaves are fleshy and often very glandular. Along our entire coast is rose-colored *Abronia umbellata* (see page 51) with rather loose heads, and from San Luis Obispo County south is a dark red fleshy species, *A. maritima*, with compact heads. The three species hybridize to some extent and a close search will usually find some plants intermediate in their characteristics.

PRICKLY-PHLOX is a low subspiny shrub related to true Phlox, but easily distinguished by having its stamens inserted at one level in the corolla-tube, while in Phlox they are at different levels. Prickly-Phlox (*Leptodactylon californicum*), plate 6, can reach to a couple of feet tall, is somewhat spreading, and may be woolly or glandular. In one form or another, it is found on dry slopes and banks below 5,000 feet from San Luis Obispo County to the Santa Ana and San Bernardino mountains. It flowers from March to June.

28

PINCUSHION CACTUS is a name given to many species of *Mammillaria*, which differ from our common prickly pears or opuntias in not having jointed stems, but in possessing tubercles that bear spines. Several species occur in southern California; on the coastal slopes there is *M. dioica*, plate 7, from the San Diego region. Others are on the deserts. The plants are mostly a few inches high, sometimes a foot, globose to cylindric, and bear scarlet conspicuous fruits. This species blooms from February to April.

PLATE 7.   PINCUSHION CACTUS

MADRONE or MADROÑO (*Arbutus Menziesii*), plate 8, is related to Manzanita (pages 19 and 76) and has similar exfoliating reddish bark and small urn-shaped flowers. It is a beautiful tree with broad crown and large shining leaves. The flowers appear from March to May and are followed by red to orange berries less than half an inch in diameter. Found on wooded slopes and in canyons below 5,000 feet, it occurs at scattered stations in southern California and is abundant in central and northern California, ranging to British Columbia. The common Strawberry Tree of cultivation is *Arbutus Unedo*, a shrub of the Mediterranean region.

PLATE 8.   MADRONE

Our common BUTTERCUP (*Ranunculus californicus*), plate 9, is a perennial herb with bright green lobed or divided leaves and shining yellow petals. In various forms it occurs widely in the state, especially on grassy and wooded slopes that are moist in the spring. It begins to bloom in February.

PLATE 9.   BUTTERCUP

PLATE 10.   VELVET CACTUS

Another coastal cactus is VELVET CACTUS (*Cereus Emoryi*), plate 10. It is conspicuous with its long more or less prostrate stems bearing ascending branches that have numerous yellow spines. It occurs on dry bluffs and cliffs from Orange County to Lower California and on Catalina and San Clemente islands. The flowers appear in May.

Related to the snapdragons and penstemons is the genus *Collinsia*, with sixteen species in California. They are annuals and have a two-lipped flower peculiar in having the middle lobe of the lower lip sunken into a structure like the keel of a boat and containing the stamens and pistil. One of our most common and most beautiful is *Collinsia heterophylla* (formerly *C. bicolor*), CHINESE HOUSES or INNOCENCE, plate 11, which ranges from almost white to quite a deep color, especially on the lower lip. It frequents shaded places below 2,500 feet, from Humboldt and Shasta counties to Lower California and flowers from March to June. See also page 72.

PLATE 11.   CHINESE HOUSES

FAREWELL-TO-SPRING or GODETIA is a large group of California annuals belonging to the Evening-Primrose Family and with four petals usually of some shade of pink or red. The one shown in plate 12 is *Clarkia amoena*, a species living on slopes and bluffs near the coast from Humboldt County to Marin County. Other related species, some with entire petals, some with divided, are found in most places west of the Sierra and south to the border. Their range of color and pattern is remarkable and they are widely cultivated, especially in northern Europe where double and other forms have been developed. The flowers in our wild plants come at or after the end of the spring rains. See also pages 42 and 56.

PLATE 12.   FAREWELL-TO-SPRING

One of our loveliest, but unfortunately rarest, shrubs is TREE-ANEMONE, plate 13 (*Carpenteria californica*). It grows to be six or many more feet tall or may even be sprawling; has leaves two to three inches long, green above and gray beneath. The flowers are a good white, about two inches across. It is found on dry slopes and ridges at 1,500 to 4,000 feet in Fresno County and begins to flower in May. It is well established in cultivation.

PLATE 13.   TREE-ANEMONE

WILD-BUCKWHEAT is a name used in general for the genus *Eriogonum* and in particular for *E. fasciculatum*, plate 14. In California there are about seventy-five species of *Eriogonum* and four forms of *E. fasciculatum*. The common form of this species extends over wide areas from Monterey County to Lower California. There is considerable variation in flower color (from deep pink to white) and in hairiness of leaves. Forming low half-woody shrubs, it is an important bee plant even though its flowers are very small. See also pages 33 and 50.

PLATE 14.   WILD-BUCKWHEAT

OWL'S-CLOVER or PAINT-BRUSH (*Orthocarpus purpurascens*), plate 15, is a low erect annual with deeply divided leaves and terminal spikes of lobed, more or less purplish bracts with narrow crimson to purplish or whitish flowers. It often occurs in great masses in pastures and on grassy hillsides and forms patches of great beauty, especially when interset with lupines and Gold-Fields. It occurs from Mendocino County to Lower California and inland to the San Joaquin Valley and the deserts. It flowers from March to May. See page 103 for another species.

PLATE 15.   OWL'S-CLOVER

31

PLATE 16. LADY-SLIPPER

LADY-SLIPPER always represents to me the acme of America's wild orchids. In California there are three species, the one shown in plate 16 is *Cypripedium californicum*, a plant one to two feet high and bearing three to ten flowers that are almost an inch long. It is found on wet rocky ledges and hillsides below 5,000 feet in Marin County, from Mendocino County to Lassen County, and in southern Oregon. It begins to bloom in May. Our other species of Lady-Slipper extends as far south as Santa Cruz and Mariposa counties and may have a more greenish or purplish lip.

PLATE 17. BUSH-SUNFLOWER

BUSH-SUNFLOWER, *Encelia californica*, plate 17, forms broad rounded clumps two to four feet tall and occurs on bluffs and in canyons near the coast from Santa Barbara County to San Diego. The showy yellow heads are two to three inches in diameter. Occasional persons are allergic to handling it. It blooms freely from February to June.

PLATE 18. MOUNTAIN-PENNYROYAL

MOUNTAIN-PENNYROYAL ranges from annual to perennial and from lavender or purplish to the red species illustrated in plate 18, namely *Monardella macrantha*. This is a creeping plant forming mats covered with red tubular flowers more than an inch long. It grows at 2,500 to 6,000 feet, in the mountains from Monterey County to Lower California, and flowers from April to July. The crushed foliage of most of the monardellas has a clean minty odor.

Wild-Buckwheat and its genus *Eriogonum* are mentioned on pages 31 and 50; the most conspicuous species in California is SAINT CATHERINE'S LACE (*Eriogonum giganteum*), plate 19. It is a coarse openly branched shrub one to several feet tall, with leaves one to three inches long and white-woolly beneath. The flower clusters are immense, in spreading cymes a foot or so across. At first the flowers are white, then turn rust on drying and make beautiful dry bouquets. Any visitor to Catalina Island should see this plant as it grows there on dry slopes.

PLATE 19. SAINT CATHERINE'S LACE

BEACH STRAWBERRY (*Fragaria chiloensis*), plate 20, has the usual trailing strawberry habit, three leaflets, and white flowers. The berry is smallish and not often seen, since male and female plants often appear in separate patches. The shining leaves and dense growth make it an attractive groundcover. Its Chilean form is one of the parents of domestic strawberries. On our immediate coast it is found primarily in sandy places from San Luis Obispo County to Alaska. Flowers begin in March.

PLATE 20. BEACH STRAWBERRY

YUCCA or OUR LORD'S CANDLE (*Yucca Whipplei*), plate 21, is known to every Californian or traveler who passes through brushy and open places in the hills and on the fans below the canyons away from the immediate coast and from Monterey and Tulare counties south. With its clumps of stiff swordlike leaves and tall stalks with innumerable white to purplish flowers, it is indeed a striking plant. Flowering from April to June, it often dominates considerable areas. It is protected by law and one no longer sees every automobile returning to the city after a Sunday excursion bear an eight foot flower stalk tied alongside. The individual flowers are an inch or more long.

PLATE 21. YUCCA

PLATE 22. GOLDEN-YARROW

GOLDEN-YARROW (*Eriophyllum conferti-florum*), plate 22, is an herb or subshrub, mostly one to two feet high, with some white wool on the stems and on the underside, at least, of the deeply lobed leaves. The yellow heads are borne in flat-topped clusters and make an attractive plant. In various forms, the species is common on brushy slopes below 8,000 feet from Mendocino, Tehama, and Calaveras counties south. The flowers begin in April.

PLATE 23. BENTHAM'S LUPINE

In the central parts of California bordering the interior valleys are two common annual blue lupines: *Lupinus Benthamii* and *L. nanus*. The former has pedicels (the short stems each of which bears a flower) about one-eighth of an inch long and linear leaflets; the latter has pedicels one-sixth to one-half inch long and often somewhat broader leaflets. Lupines (see page 13) are characterized by mostly having their leaves divided into several leaflets in a palmate fashion, that is, with the leaflets coming out at a common point. In BENTHAM'S LUPINE, plate 23, the flowers are about half an inch long and appear from March to May.

PLATE 24. CALIFORNIA BUCKEYE

The CALIFORNIA BUCKEYE (*Aesculus californica*), plate 24, is a large bush or a small tree with a broad round top. At the onset of the dry season it sheds its leaves and its bare branches are often quite conspicuous with their large brown "horse-chestnuts." But when the rains come and leaves appear, there emerge dense white spikes almost a foot long of white or pale rose flowers with long stamens. The effect is striking. The Buckeye grows on dry slopes and in canyons below 4,000 feet from Siskiyou and Shasta counties to northern Los Angeles County and flowers in May and June. The Indians ground up the seed to use in stupefying fish; the flowers are said to be quite poisonous to bees.

GILIA or BIRD'S EYES (*Gilia tricolor*), plate 25, is a handsome annual growing on open grassy slopes and plains below 2,000 feet in the Coast Ranges from Humboldt County to San Benito County, in the Central Valley, and along the base of the Sierra Nevada, especially northward. Its color ranges from pale to deep blue-violet and its beauty is enhanced by its five pairs of purple spots. Growing to about one foot high, it is one of the most attractive members of its group, flowering from very early spring.

PLATE 25.   BIRD'S EYES

Some lupines are of quite long duration and form true shrubs with woody trunks. One example is BUSH LUPINE (*Lupinus albifrons*), plate 26, found in sandy and rocky places below 5,000 feet and from Humboldt and Shasta counties south. It has various forms that are only technically separable, but is in general characterized by very long spikes of flowers and the middle upper petal more or less hairy on the back. Another common shrubby species is *Lupinus arboreus* of the immediate coast, with shorter spikes of more rounded flowers ranging in color from blue to yellow and not hairy on the back. Both begin to bloom in March.

PLATE 26.   BUSH LUPINE

In the Coast Ranges from Contra Costa County south and over into the Mojave Desert is a common annual White Daisy or WHITE LAYIA (*Layia glandulosa*), plate 27, covering sandy areas in great profusion. What appears to be a single flower is really a head of flowers as in the Sunflower, with small central tubular ones on the disk and marginal petallike ones at the edge. The plant is mostly about a foot high and begins to bloom in March. See also page 85.

PLATE 27.   WHITE LAYIA

PLATE 28. GROUND-PINK

Linanthus is mentioned in other sections (pages 21 and 89) as a group related to *Phlox* and *Gilia*, but with leaves palmately divided into linear lobes. One of the most interesting of the species is GROUND-PINK or FRINGED-PINK (*Linanthus dianthiflorus*), plate 28, which hugs the ground rather closely and forms small plants with exceedingly slender stems and pink to lilac or white flowers. It is common in open sandy places below 4,000 feet from Santa Barbara to San Diego and begins to bloom shortly after Christmas.

PLATE 29. MALVA ROSA

TREE-MALLOW or MALVA ROSA (*Lavatera assurgentiflora*), plate 29, is a large bushy shrub with five- to seven-lobed leaves to about six inches across. The flowers are two to three inches across or more. Originally from the Channel Islands off our coast, Malva Rosa became popular among the early Mexican settlers and has escaped from cultivation on the mainland. It blooms from March to November.

PLATE 30. WILD-CANTERBURY-BELL

After burns and in other disturbed places a common spring wildflower is WILD-CANTERBURY-BELL (see page 70). The one figured in plate 30 is *Phacelia minor*, with a deep truly bell-shaped corolla from about one-half to more than an inch long and with glandular ovate leaves. It occurs from Los Angeles County to Lower California and blooms from March to June.

The monkey-flowers in California are very numerous and of many colors (compare pages 62, 85, and 103). Several are quite woody and make shrubs up to about three feet high. All are characterized by their angled calyx and two-lipped corolla. One of the more conspicuous is BUSH MONKEY-FLOWER (*Mimulus longiflorus* and varieties), plate 31, which is common on dry slopes below 5,000 feet in central and southern parts of the state. A related, less hairy species (*M. bifidus*) is in our more northern counties. Flowering from March to August, these shrubs can be cut back and a second growth encouraged for later in the season.

PLATE 31. BUSH MONKEY-FLOWER

PLATE 32. SUN CUP

A more localized species and usually away from the immediate beach is SUN CUP (*Oenothera bistorta*), plate 32, which in various forms is found from Los Angeles and Kern counties south. It is annual and inhabits mostly sandy or dry disturbed places and begins to flower in January. The flowers often have dark spots near their center.

COREOPSIS (*C. gigantea*), plate 33, is a stout often unbranched shrub two to several feet high, which during the dry season greatly resembles a dead broomstick. When the winter rains come, a tuft of bright green finely divided leaves appears at each tip and is soon followed by large yellow daisylike heads of flowers. It is found on rocky sea cliffs and exposed dunes from Los Angeles County to San Luis Obispo County and on the Channel Islands. It flowers from March to May.

PLATE 33. COREOPSIS

PLATE 34.   BEACH-PRIMROSE

The evening-primroses have many different types and forms, some as the large-flowered ones (page 39) opening toward evening and having four linear lobes to the stigma, others with an undivided spherical stigma. In this category belong such day-bloomers as the BEACH-PRIMROSE (*Oenothera cheiranthifolia*), plate 34, a prostrate perennial of sandy beaches and with grayish foliage. It is found from Oregon to Lower California, growing more woody in the southern part of its range. It is in bloom much of the year.

PLATE 35.   WOOLLY BLUE CURLS

WOOLLY BLUE CURLS or ROMERO (*Trichostema lanatum* [and its less woolly relative *T. Parishii*]), plate 35, are low rounded shrubs with narrow leaves and long spikes of blue flowers enclosed in woolly calyces with blue, pink, or almost white hairs. Being in the Mint Family, they are quite aromatic. They are found in dry bushy places mostly below 5,000 feet and from Monterey County to Lower California. Flowering is primarily in the spring. See page 70.

PLATE 36.   WESTERN AZALEA

WESTERN AZALEA (*Rhododendron occidentale*), plate 36, is a loosely branched deciduous shrub three to fourteen feet tall. The leaves are one to three inches long. The flowers are usually white with a yellowish blotch, or they may have a pink tinge. Western Azalea is found in damp spots and along streams below 7,500 feet, in the pine belt in the mountains of San Diego and Riverside counties and in the Sierra Nevada, but extending to very low elevations in the Coast Ranges from Santa Cruz County north to Oregon. It flowers from April to August. For our other species of *Rhododendron* see page 85.

Evening-Primrose (see page 100) is a name applied to various plants with white or yellow flowers that open about sundown and close when the sun of the next day gets bright. The white one, *Oenothera deltoides*, plate 37, is often called DESERT-PRIMROSE, since it is most common on the Mojave Desert. However, in forms varying as to type of hair and shape of buds, it extends through the San Joaquin Valley to the sand dunes near Antioch. Mostly an annual, a foot or less high, with coarsely toothed or lobed leaves, it forms quite showy masses in sandy open places. The stigma has four linear lobes. Flowers begin in early spring.

PLATE 37.   DESERT-PRIMROSE

LIVE-FOREVER is a name applied to a rather large group of succulents of the genus *Dudleya* (see pages 14 and 88). They are sometimes called Hen-and-Chickens because of the habit of some to proliferate and send out new young plants from the base. One of the showiest is a coastal species extending from San Luis Obispo County south and found on rocky cliffs and in canyons. This largest of our native live-forevers is *Dudleya pulverulenta*, plate 38, growing to over a foot high and with the broad leaves covered with a dense white mealy powder. The petals are deep red, a half-inch or more long. Flowering is from May to July.

PLATE 38.   LIVE-FOREVER

SCARLET LARKSPUR (*Delphinium cardinale*), plate 39, seems out of place in a genus with so many blue-flowered species (see pages 67 and 84), but some are rose-colored and others scarlet, such as *Delphinium nudicaule*, a red-flowered type from Monterey County north. The one here illustrated occurs from Monterey County south. It reaches from three to six feet in height. Because its basal leaves wither early, by the time it comes into flower it is mostly a long leafless open panicle of spurred red flowers. It occurs in open dry places among brush and woods and begins to bloom in May.

PLATE 39.   SCARLET LARKSPUR

39

PLATE 40.   PRICKLY POPPY

With white flowers like those of Matilija Poppy, but more crinkled and smaller, is PRICKLY POPPY (*Argemone munita*), plate 40. Growing two to four feet high, the plant is covered throughout with stiff spines. It is found in various forms in much of California and begins to bloom in March.

CAMAS (*Camassia Quamash*), plate 41, is another species with several forms, but can be treated here as a unit. It has bulbs up to about an inch thick, leaves one to one and one-half feet long, and a flowering stalk to about two feet high. The deep blue-violet flowers are highly attractive, especially when a meadow of many acres may be almost a solid sheet of color. Found in wet meadows from Marin and El Dorado counties north, it flowers from May to July. The roasted bulbs were eaten by the Indians.

PLATE 41.   CAMAS

WIND POPPY or FLAMING POPPY (*Stylomecon heterophylla*), plate 42, is an erect, usually several-stemmed annual with pinnately lobed leaves and terminal orangered flowers with purplish spots. The petals are one-half to almost an inch long. This species is occasional on grassy and brushy slopes below 4,000 feet from Lake County through the Coast Ranges to northern Lower California and in southern Sierra Nevada. The flowers come in April and May. It very much resembles *Papaver californicum* (see page 47) of the same common name, but that Poppy has the stigma flat and immediately on the pod, while *Stylomecon* has it more spherical and elevated above the pod.

PLATE 42.   WIND POPPY

MATILIJA POPPY (*Romneya Coulteri*), plate 43, is one of our most elegant flowers. Growing on rather woody stems three to seven feet high, the several blossoms have crinkled petals two to four inches long with clear yellow stamens. The leaves are grayish-green and parted or divided into three to five main divisions. The sepals are smooth, while in the related *Romneya trichocalyx* they have long stiff hairs. Matilija Poppy is found in dry washes and canyons from Ventura County to Lower California and blooms in May and June.

PLATE 43. MATILIJA POPPY

PRICKLY-PEAR has many forms and colors and is found in much of California, especially on dry slopes and fans and in canyons. Flowers may be yellow or orange or even more reddish. The stem consists of flattened joints with so-called areoles from which arise spines and stiff prickly hairs or glochids. The fruit is usually fleshy and may have a desirable flavor. One must be certain first to remove with considerable care all the glochids from the outer surface, otherwise his tongue suffers. These prickly-pears are opuntias; the one shown in plate 44 is *Opuntia occidentalis* var. *Vaseyi* of interior southern California. Prickly-pears flower mostly in May and June. See also page 27.

PLATE 44. PRICKLY-PEAR

CALIFORNIA-LILAC or CEANOTHUS, plate 45, is found in brushy places from one end of California to the other and ranges through a remarkable series of forms from deep blue to lavender and light blue and finally white. Occurring in brushy and wooded places from the pine belt to sea level, California-lilacs add more to the beauty of our spring landscape than almost any other plants. They vary from prostrate matted forms to tall treelike shrubs. Only the expert can distinguish the many species. See also page 18.

PLATE 45. CALIFORNIA-LILAC

41

PLATE 46. FREMONTIA

California Slippery-Elm is an old name, but more common now is FREMONTIA or FLANNEL BUSH (*Fremontia californica*), plate 46, for a large open shrub that varies greatly in leaf- and flower-size. It is one of America's showiest shrubs, with open flat flowers one to two or more inches in diameter and of a beautiful clear yellow. It is found in various forms in the Sierran foothills and inner Coast Ranges from Lake and Napa counties south, as well as in scattered stations in southern California and Arizona. Flowering is in April and May. A related local species on the Mexican border is *F. mexicana*, much used in cultivation and with some red in the flowers especially in age.

COLUMBINE (compare page 46) is found widely spread over the northern hemisphere and ranges from yellow or almost white to red, red and yellow, and blue to purple. In California we have a common species of the red and yellow type in moist places from sea level to high elevations, but the one illustrated in plate 47, *Aquilegia eximia*, is more local. It is remarkable in having the blade of the petal entirely lacking and only the spur is left. It occurs in seeps on serpentine in the Coast Ranges from Mendocino County to Ventura County and begins flowering in May.

PLATE 47. COLUMBINE

PLATE 48. GODETIA

Another FAREWELL-TO-SPRING or GODETIA (see pages 30 and 56) is *Clarkia Bottae*, plate 48. It closely resembles several other species that are widely distributed in California and flower at the beginning of the dry season. This particular species is from Monterey County.

FLOWERS ROSE TO PURPLISH-RED
OR BROWN

Section Two

In England a popular garden plant is the Fritillary of which there are some quite handsome Eurasian species. On the Pacific Coast we have several that may not be so beautiful, but are of some interest. One of these is CHOCOLATE-LILY (*Fritillaria biflora*), figure 59, with dark brown or greenish-purple flowers an inch or more long. It grows in grassy places from Mendocino County to San Diego County and begins to flower in February. Other species may be scarlet, orange, spotted. Some have peculiar little bulblets like grains of rice about the main bulb.

FIGURE 59.   CHOCOLATE-LILY

BROWNIES is the name given to another lilylike plant, *Scoliopus Bigelovii*, figure 60, with greenish flowers having reddish veins; the broad leaves have purplish mottling. This species is found on shaded slopes from Humboldt County to Santa Cruz County.

In damp woods from Del Norte County to Monterey County is CLINTONIA (*C. Andrewsiana*), figure 61, with rose-purple flowers and beautiful dark blue berries.

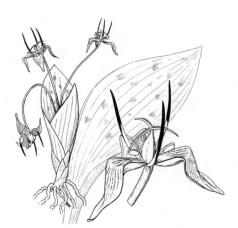

FIGURE 60.   BROWNIES

FIGURE 61.   CLINTONIA

FIGURE 62.   PEONY

FIGURE 63.   COLUMBINE

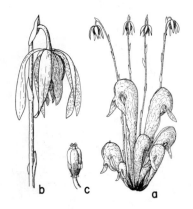

FIGURE 64.   CALIFORNIA PITCHER-PLANT

Many Californians envy their Eastern friends who can raise such lovely peonies in their gardens. They do not realize that California has two native species, but after seeing these they may still envy the northerners who grow peonies. To me our PEONY (*Paeonia Brownii*), figure 62, is an interesting rather than beautiful plant. The drooping flower does not open up enough to show what beauty it might have. Our two species are much alike: *P. Brownii* occurring from Santa Clara and Tuolumne counties north and *P. californica* from Monterey County south. The foliage is handsome, somewhat fleshy; the petals are maroon to blackish-red at the center. In the same Buttercup Family is the COLUMBINE, an old favorite both in the garden and wild. The flowers of many species are nodding and red to yellow; among these is *Aquilegia formosa*, figure 63, which grows in somewhat different forms from one end of the state to the other and in damp places from sea level into high mountains. The petals are short, each ending in a spur that contains nectar. See also page 42.

One of our oddest plants is the CALIFORNIA PITCHER-PLANT (*Darlingtonia californica*), figure 64, of marshy and boggy places from Nevada and Trinity counties north to Oregon. Each rather tubular leaf ends above in an arched hood with two hanging lobes and is a very successful trap for catching and digesting insects. The petals are dark purple, an inch or more long; the sepals are yellow-green with purplish lines. The plants are one to two feet high.

This book shows some white and some yellow poppies (see pages 40, 41, 82, and 94) and here illustrated is a red POPPY (*Papaver californicum*), figure 65, a slender annual one to two feet tall, with brick red flowers an inch or so in diameter. The seed pod is short and flattened at the top. The plant occurs on burns and in recently disturbed places from Marin County to San Diego County. Associated with it, but getting into the foothills of the Sierra Nevada and north to Lake County, is a similar Poppy (*Stylomecon heterophylla*) in which the seed pod ends in a persistent style with a small headlike stigma at the apex. (See page 40).

Poppies mostly have four petals; so do mustards. One of the latter, CAULANTHUS COULTERI, figure 66, is a hairy erect annual one to almost three feet tall, often with purplish sepals and crisped white petals with purple veins. The seed pod shown at the left is an inch or more long. This plant in one form or another is found from Alameda and Madera counties to Los Angeles County.

A big group in the Mustard Family is the ROCK-CRESS or ARABIS. In figure 67 is shown *Arabis blepharophylla*, a perennial less than a foot high, coarse-hairy, with rose-purple petals about half an inch long. It is found in rocky places from Sonoma County to Santa Cruz County. See also figure 68.

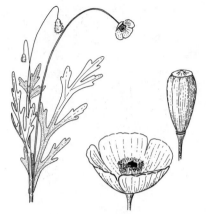

FIGURE 65.   POPPY

FIGURE 66.   CAULANTHUS

FIGURE 67.   ROCK-CRESS

FIGURE 68.   BREWER'S ROCK-CRESS

FIGURE 69.   FEW-FLOWERED ROCK-CRESS

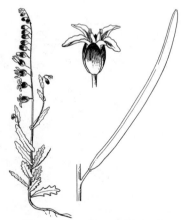

FIGURE 70.   JEWEL FLOWER

Another species of Mustard is BREWER'S ROCK-CRESS (*Arabis Breweri*), figure 68, of about the same height. It has red-purple to pink flowers, about half an inch across, and pods one to three inches long. It is found in rocky places from Monterey and Yuba counties to Oregon. Closely related forms are in the San Bernardino Mountains and in Butte County.

The FEW-FLOWERED ROCK-CRESS (*Arabis sparsiflora*), figure 69, runs through a number of variations. In general it is hairy only below, one to three feet high, with pink to purple flowers, and is found on dry slopes up to 9,000 feet. In its various forms it is distributed through most of our area.

Close to Caulanthus, figure 66, is the JEWEL FLOWER (*Streptanthus glandulosus*), figure 70, also of the Mustard Family. It is exceedingly variable and has a number of named varieties. In general, it is stiff-hairy, at least below, is mostly one to two feet high and has purplish flowers or white with purple veins. The seed pods are more or less ascending to erect, sometimes recurved, often two or three inches long. It is found on dry often loose disturbed soil or on rocky ridges, from Mendocino County to Santa Clara County.

The SEA-ROCKET (*Cakile maritima*), figure 71, is another member of the Mustard Family that has become naturalized from Europe, growing on beach sands from Mendocino County to Monterey County. The stems tend to lie on the sand or ascend somewhat; flowers are pinkish or purplish, scarcely half an inch long. A native species, *C. edentula* subspecies *californica*, with less deeply divided leaves and smaller flowers, occurs on beaches from San Diego County to British Columbia.

TAMARISK (*Tamarix pentandra*), figure 72, is a loosely branched shrub with very slender twigs and minute appressed scaly leaves. The pinkish or almost white flowers are small and very numerous. Introduced from Europe, Tamarisk has become common in low places and along water courses.

TURKISH RUGGING (*Chorizanthe staticoides*), figure 73, is an erect annual about five to eight inches high, with basal leaves and minute rose-pink, rather persistent, dry flowers in dense clusters. It grows on dry slopes and flats from Monterey County to San Diego.

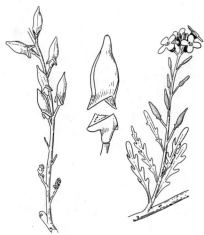

FIGURE 71. SEA-ROCKET

FIGURE 72. TAMARISK

FIGURE 73. TURKISH RUGGING

FIGURE 74.   CHORIZANTHE MEMBRANACEA

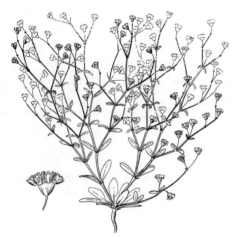

FIGURE 75.   WILD-BUCKWHEAT

FIGURE 76.   WILD-RHUBARB

Related to TURKISH RUGGING (figure 73), but with longer stems and woollier flowers is CHORIZANTHE MEMBRANACEA, figure 74, from dry rocky places and ranging from Mendocino and Siskiyou counties to Ventura and Kern counties. In these chorizanthes the flowers have six petallike parts. They are related to WILD-BUCKWHEAT, the genus *Eriogonum*, with about seventy-five species in California. The buckwheats differ from the chorizanthes in having a cup-shaped or cylindrical involucre around the cluster of flowers instead of separate bracts. Among the buckwheats are annuals, like *Eriogonum gracillimum*, figure 75, which may have leaves running up the stems or entirely in a basal rosette. They are found everywhere in California, especially in dry and disturbed places. The one illustrated is common on sandy plains below 3,500 feet, in the inner Coast Ranges from Monterey and Merced counties south through the interior of southern California. The flowers are rose tipped with white. See also pages 31 and 33.

The same family has plants like the garden Rhubarb (*Rheum*) and Dock (*Rumex*). A species of the latter is WILD-RHUBARB or CANAIGRE (*Rumex hymenose-palus*), figure 76, a heavy perennial two to three feet tall, stout, with large fleshy leaves and terminal panicles up to a foot long of small pinkish flowers. They are followed by winged fruits about a half-inch long. Wild-Rhubarb is found in dry sandy places from Kern and San Luis Obispo counties south. It has been used for food. Being astringent, it has also been employed for tanning.

Saltbush is a name for a large group of shrubs or herbs, mostly covered with a scurfy grayish coat of inflated hairs and with very small flowers. The seed is enclosed between a pair of bracts. The plants are usually found where the soil is alkaline or saline, as along beaches and in undrained valleys. AUSTRALIAN SALTBUSH (*Atriplex semibaccata*), figure 77, is a prostrate perennial with reddish fleshy fruiting bracts and has become naturalized from Australia. It is abundant in waste places from San Luis Obispo County to Lower California and is in the San Joaquin and Imperial valleys. See also pages 26 and 96.

In the Four-o'Clock Family California has quite a representation. Outstanding are the plants called SAND-VERBENA, found in many sandy places. They are trailing plants, one of the more common being *Abronia umbellata*, figure 78, a perennial with stems to about one yard long and with mostly rose-colored flowers in dense clusters inclosed by an involucre of several bracts. It is found on the coastal strand from Del Norte County to San Diego. With it may grow a stouter viscid plant with smaller darker flowers, *Abronia maritima*. For a yellow species see page 28.

The Catchfly or Campion, sometimes called INDIAN-PINK, is the genus *Silene* of the Pink Family. The species figured is *S. californica*, figure 79, with crimson flowers about an inch in diameter. It is found in brushy or wooded places from Los Angeles and Kern counties to Oregon. A similar species with narrower leaves is *S. laciniata* from Santa Cruz County to Mexico. In pioneer days it was used as a tea for aches, sprains, and sores.

FIGURE 77. AUSTRALIAN SALTBUSH

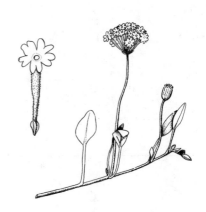

FIGURE 78. SAND-VERBENA

FIGURE 79. INDIAN-PINK

FIGURE 80. WILD-GINGER

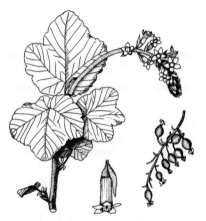

FIGURE 81. WILD CURRANT

FIGURE 82. SEA-FIG

WILD-GINGER is a group of perennial herbs with creeping stems and mostly heart-shaped leaves that have a spicy odor when crushed. California has three species; one, *Asarum caudatum,* figure 80, being found in deep shade from Santa Cruz Mountains northward. The brownish-purple flowers have sepals only and no petals and may be almost hidden in among the leaves.

Later in this book are shown yellow-flowered currants (page 96); here is a WILD CURRANT with pink flowers, *Ribes malvaceum,* figure 81. It is sometimes called Chaparral Currant, since it is a common chaparral species, found from Tehama and Marin counties to Orange County. It is quite glandular and aromatic. Another species with even redder flowers is *Ribes sanguineum,* the Red Flowering Currant, that ranges north to British Columbia. It is widely cultivated in western Europe. Both of these species have black or purple-black berries, but so glandular as not to be very useful. See page 75.

Among the succulent plants of California are Ice Plant, Sea-Fig, Hottentot-Fig, and the like, all of which are in the genus *Mesembryanthemum.* The one shown here, figure 82, is *M. chilense,* the SEA-FIG, so-called because it is found along our sea-coast and has rather a fleshy fruit. The flowers are magenta and one to two inches in diameter. See also page 77.

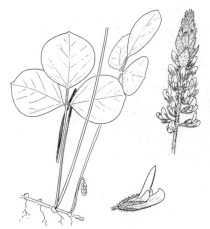

FIGURE 83.   PSORALEA

PSORALEA is the name for a group of perennial herbs with numerous dots or glands and heavy-scented foliage. The leaflets are three and the flowers are pea-like. One species, *Psoralea orbicularis*, figure 83, has creeping stems with flower stalks one to two feet tall; the reddish-purple flower is about half an inch long. It is found in moist places below 5,000 feet through most of our area. Other species, some much taller and with erect stems, are in dryer places.

An attractive little shrub is CHAPARRAL-PEA (*Pickeringia montana*), figure 84, which grows from one to six feet tall, is somewhat spinescent, and has the leaves divided into one to three leaflets. The purple flowers are more than half an inch long, the pod one to two inches. It is found on dry slopes and ridges below 5,000 feet from Mendocino and Butte counties to San Diego.

Mallow, Hollyhock, Cotton, and Hibiscus are familiar names to a plant lover. Their family, the Mallow Family, is well represented in California. A large group of fairly coarse, often somewhat woody shrubs, is *Malacothamnus*, with FALSE-MALLOW (*M. densiflorus*), figure 85, as an example. It has the scurfy pubescence characteristic of many mallows, grows to be about six feet tall, and has rose-pink flowers almost an inch across. It is found on dry slopes in Riverside and San Diego counties and is closely matched by similar species farther north.

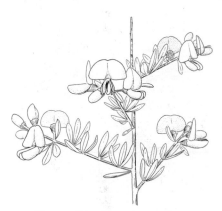

FIGURE 84.   CHAPARRAL-PEA

FIGURE 85.   FALSE-MALLOW

FIGURE 86.   CHECKER

FIGURE 87.   MALVASTRUM PARRYI

FIGURE 88.   REDWOOD-SORREL

CHECKER or WILD HOLLYHOCK is another member of the Mallow family and is the common name for *Sidalcea malvaeflora*, figure 86. This plant is most variable and is represented by innumerable forms from grassy, sometimes damp, places of much of California. It is a perennial with stems one to two or more feet high and terminal racemes of pink flowers about an inch across.

Another Mallow is an annual, MALVAS-TRUM PARRYI, figure 87, largely at the edge of the San Joaquin Valley from Alameda County to Ventura County. The petals are pinkish-lavender to purplish.

Oxalis or Wood-Sorrel is discussed under yellow-flowered plants (page 99). A species with pinkish flowers, often veined purple, is the REDWOOD-SORREL (*Oxalis oregana*), figure 88. It is found mostly in shade from Monterey County north to Washington. It has wiry scaly rootstocks and is stemless above the ground.

The Geranium Family has its fruit sepa-
rating into five parts when mature; these
pull apart at the base and coil into twisted
tails. Such a plant is FILAREE or CLOCKS
(*Erodium cicutarium*), figure 89. It is an
annual from Europe that has become
widely naturalized in California and is
important for forage for sheep and cattle.
The flowers are rose-lavender.

A native of America is CRANESBILL (*Geranium carolinianum*), figure 90, also a
spring annual and widely distributed. The
illustration shows how the fruit pulls apart.

A very different group is the Evening-
Primrose Family. The flower is on the plan
of four: four sepals, four petals, four or
eight stamens. The young capsule in which
the seeds are produced is beneath the
flower and can be seen from the outside.
BOISDUVALIA is an example (*B. densiflora*),
figure 91, widely spread in the state and
with rose-purple notched petals. Resembling it, but easily recognized by the tuft
of hairs on each seed, is the WILLOW HERB
or Fireweed. Some species are annual, but
most are perennial; many are found in wet
places.

FIGURE 89.   FILAREE

FIGURE 90.   CRANESBILL

FIGURE 91.   BOISDUVALIA

FIGURE 92.   WILLOW HERB

FIGURE 93.   CLARKIA

FIGURE 94.   GODETIA

Epilobium Watsoni, figure 92, is one of the coastal willow herbs which, in one form or another, grows from San Luis Obispo County to Oregon. The petals are red-purple, about one-third inch long.

In the same family is FAREWELL-TO-SPRING or GODETIA, now placed in the genus *Clarkia,* named for Captain William Clark of the Lewis and Clark Expedition to the Pacific Northwest early in the nineteenth century. *Clarkia unguiculata,* figure 93, with its narrow petal bases is a true CLARKIA, found from San Diego to Mendocino and Butte counties. *Clarkia purpurea* subspecies *viminea,* figure 94, with broader petal bases is of the GODETIA type. It and related species are found in grassy and brushy or wooded places in most of California west of the Sierra Nevada and of the more southern mountains. They run through many shades of rose and lavender and may or may not have central blotches on the petals. They are often called Farewell-to-Spring, since they are among the last annuals of our California spring flowering season. See color plates on pages 30 and 42.

MILKWORT is the name given to rather a cosmopolitan group of plants that were at one time supposed to increase the flow of milk. They have a pealike flower, but with a very different make-up from the pea, the large colored parts being mostly sepals rather than petals. *Polygala californica*, figure 95, one of our two most common species, is up to about one foot in height and generally has rose-colored flowers about half an inch long. It is found in brush and open woods on rocky ridges and slopes from San Luis Obispo County north, while a closely related one, *P. cornuta*, ranges the length of the state.

FIGURE 95.   MILKWORT

The Carrot Family is mentioned on page 101. Many members of the family have white or yellow flowers, but the PURPLE SANICLE (*Sanicula bipinnatifida*), figure 96, may have them quite dark. The Sanicle has several flowers in a headlike cluster and a number of these clusters arise from near a common point. The flowers are minute. This species is found in rather open places from Lower California to British Columbia. An uncommon but most attractive little perennial herb is CYCLADENIA (*C. humilis* var. *venusta*), figure 97. It grows usually in patches in rocky places at 3,000 to 9,000 feet, from Los Angeles County to extreme northern California. The rose-purple flower is more than half an inch long and is followed by seed pods over two inches long.

FIGURE 96.   PURPLE SANICLE

FIGURE 97.   CYCLADENIA

FIGURE 98.   MILKWEED

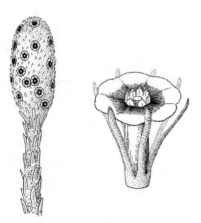

FIGURE 99.   PHOLISMA

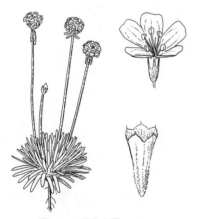

FIGURE 100.   THRIFT

Distantly related to the foregoing Cycladenia, and with milky sap as in that plant, is MILKWEED, of which California has a dozen species. One of these is *Asclepias californica*, figure 98, a soft-woolly perennial up to two feet high with purplish or dark maroon flowers. It is frequent on dry slopes through much of California west of the Sierra Nevada and more southern ranges. Some of the milkweeds were used in early days for healing cuts and for rheumatism.

Along the sandy beaches of California are found two little herbs, both quite attractive and interesting. One is PHOLISMA (*P. arenarium*), figure 99, which is parasitic on roots of other plants and has no chlorophyll of its own. The flower is about an eighth of an inch broad and is purplish with a white border. Pholisma is found from San Luis Obispo County south.

The other herb is THRIFT (*Armeria maritima* var. *californica*), figure 100, a tufted stemless plant with rose-pink flowers in small dense heads. It is in coastal sandy places and on sea bluffs from San Luis Obispo County north.

Everyone knows the cultivated Cycla-
men, a European herb with nodding
flowers and reflexed petals. In America we
have a related plant, the SHOOTING-STAR
(*Dodecatheon*), figure 101. In the lower
elevations of California are two or three
quite similar species, often becoming a foot
or so tall and forming great patches in
grassy places through much of the state.
They tend to be lavender to rose, often
with a maroon band, and to have a delicate
fragrance. When we think of Gentian, we
think of blue, but there are green gentians
(page 101) and in the same family is a
group with rose or pink flowers, CENTAURY
or CANCHALAGUA.

Figure 102 (*Centaurium venustum*)
shows our common species, an annual a
foot or so tall, with flowers the size of a
silver quarter or half-dollar and with peculiar twisted stamens. It is found on dry
slopes and flats from Butte and Shasta
counties to San Diego.

One of the more conspicuous families in
California is the Phlox Family, with the
petals united and with three stigma-lobes.
Among the spring bloomers is POLEMO-
NIUM CARNEUM, figure 103, a perennial one
to two feet high, with thirteen to twenty-
one leaflets and a purple to pinkish flower
one-half to one inch in diameter. It is
found in grassy and brushy places from
San Mateo northward.

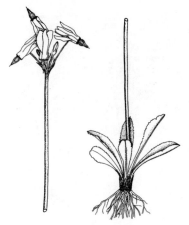

FIGURE 101.   SHOOTING-STAR

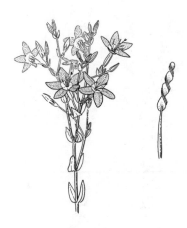

FIGURE 102.   CANCHALAGUA

FIGURE 103.   POLEMONIUM

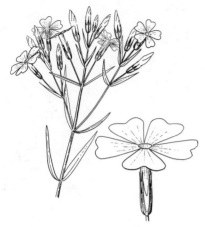

FIGURE 104. PHLOX

FIGURE 105. MINT-LEAVED VERBENA

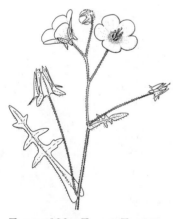

FIGURE 106. FIESTA FLOWER

In PHLOX the corolla-lobes flare at right angles to the tube and when the flower is held up to the light, the stamens can be seen at different heights in the tube. *Phlox speciosa* subspecies *occidentalis*, figure 104, is about a foot high, somewhat woody at the base, and with bright pink flowers up to an inch in diameter. It grows on wooded slopes and rocky places, at 1,500 to 7,000 feet, from Sonoma County north and along the base of the Sierra from Fresno County north.

Verbena is a name familiar to the gardener. In addition to the cultivated species is a series of weedy plants found in waste places and near streams, some native, some introduced. Some have small purple flowers in clusters that elongate after flowering; others are blue. As an example is illustrated the MINT-LEAVED VERBENA (*V. menthaefolia*), figure 105, of southern California.

Related to Baby-Blue-Eyes and like it in having bracts between the sepals, but with purple rather than blue flowers, is FIESTA FLOWER (*Pholistoma auritum*), figure 106. It is a sprawling annual with recurved prickles on the stems so that it clings to clothing. The flowers are up to almost an inch in diameter. The plant grows in deep canyons and on shaded slopes from Lake and Calaveras counties south and blooms from March to May.

Members of the Mint Family tend to have four-cornered stems and are aromatic because of their essential oils. The flowers are two-lipped and, instead of a capsule or seed pod, each flower usually forms four one-seeded nutlets. In this family we find HEDGE-NETTLE (*Stachys bullata*), figure 107, with purplish flowers more than half an inch long. It is a stiff-hairy perennial and found on dryish slopes and in canyons from San Francisco to Orange County.

Another of the mints is PENNYROYAL, the one shown in figure 108 being *Monardella lanceolata*. It is an erect annual one to two feet tall, with a pleasant odor to the crushed foliage. The flowers are about half an inch long, rose-purple or paler. It is found in dry places from Shasta County to Kern County and from San Luis Obispo County to San Diego. Other annual species occur elsewhere and some species are perennial, either bushy or creeping. See also page 32.

The PAINT-BRUSH is familiar to every westerner who knows the out-of-doors. It has small or rather narrow flowers in dense terminal clusters among brightly colored, usually scarlet bracts. The genus is *Castilleja*, named for a Spanish botanist. Most species are usually parasitic on roots of other plants. *Castilleja affinis*, figure 109, is common on dry brushy or wooded slopes from Sonoma and Napa counties to Lower California.

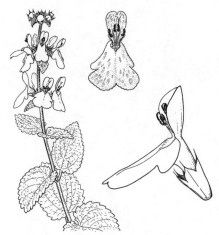

FIGURE 107.   HEDGE-NETTLE

FIGURE 108.   PENNYROYAL

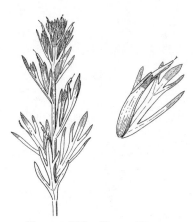

FIGURE 109.   PAINT-BRUSH

FIGURE 110.   PAINT-BRUSH

FIGURE 111.   SCARLET BUGLER

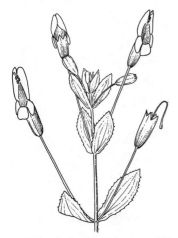

FIGURE 112.   SCARLET MONKEY-FLOWER

Contrasted to *C. affinis* of figure 109 is another PAINT-BRUSH (*C. latifolia*), figure 110, a more coastal species with broader bracts and stubbier flowers. It ranges in various forms from Humboldt County to Monterey County.

In the same family is Beard-Tongue or Penstemon, so called because the fifth stamen is sterile and often flattened and bearded like an elongate hairy tongue. Our most common red species is SCARLET BUGLER (*P. centranthifolius*), figure 111, a perennial with several stems one to three feet high and found in dry disturbed places from Lake County south. The tubular scarlet flower is about an inch long. For other penstemons see pages 72 and 90.

Monkey-Flower is a name given to a large group, *Mimulus*, with almost eighty species in California. Many are annual, many perennial; colors run from yellow to scarlet to purple. Our SCARLET MONKEY-FLOWER (*M. cardinalis*), figure 112, is a freely branched viscid perennial, frequent on stream banks and in wet places of California coastal and montane areas. It is in cultivation in Europe, where it has had a large number of color forms developed for garden use. Other monkey-flowers are shown on pages 37, 85, and 103.

In the same family with the Paint-Brush and Monkey-Flower is the Lousewort; like Paint-Brush, it too is a partial parasite on roots of other plants. Our most common spring species is also called INDIAN WARRIOR (*Pedicularis densiflora*), figure 113. It can become almost two feet high and bears deep purple-red flowers about an inch long. Found on dry slopes in chaparral and woodland, it ranges the length of the state.

The genus that gives its name to the family (Scrophulariaceae) but not nearly so showy as paint-brushes and monkey-flowers and penstemons, is FIGWORT (*Scrophularia*), a large group particularly in Eurasia. Our common species in California (*S. californica*), figure 114, has smallish red-brown to maroon flowers in large open clusters. The plants attain a height of several feet and are found in rather open places from near the coast to the interior and from Oregon to San Diego. It is a plant highly attractive to bees.

This same large family is here represented by two other illustrations. One, the SNAPDRAGON (*Antirrhinum multiflorum*), figure 115, is a stout, widely branched, exceedingly viscid plant with rose-red flowers having a white or cream lower lip or palate. It is found on dry slopes below 4,000 feet in the central Sierran foothills and from Santa Clara County to San Bernardino County. Other snapdragons, some with reddish, some with white flowers, are found in California (see page 24).

FIGURE 113. INDIAN WARRIOR

FIGURE 114. FIGWORT

FIGURE 115. SNAPDRAGON

FIGURE 116. FOXGLOVE

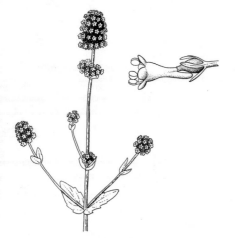

FIGURE 117. PLECTRITIS

The FOXGLOVE (*Digitalis purpurea*), fig. 116, is a native of Europe, but has established itself freely in more or less shaded places near the coast from Santa Barbara County to British Columbia. Its purple to white flowers are up to two inches long. Compare with figures 113, 114, and 115.

On grassy and wooded slopes in much of the state there appear in great numbers small spring annuals from an inch or so up to a foot high and with minute usually small, pinkish, two-lipped flowers. They are so abundant that they cannot escape notice, but no common name seems available. They belong to the genus PLECTRITIS; one species (*P. macrocera*) is shown in figure 117. They are in the Valerian Family, which may be known to some of you readers with a knowledge of medicinal plants and of the Garden Valerian (*Centranthus ruber*).

FLOWERS BLUE TO VIOLET

Section Three

Pacific Coast irises are an unusually attractive group, with rather narrow leaves
and with flowers running through a wide
range of colors, partly because of hybridi-
zation. The most common and most robust
IRIS is *I. Douglasiana*, figure 118, forming
large clumps of tough fibrous leaves with
pinkish or reddish bases. The flower stalks
attain a height of one or two feet; petals
are two to three inches long and vary from
pale lavender to blue or deep red-purple.
The plants are abundant on grassy slopes
and in open places, from Santa Barbara
County to Oregon.

FIGURE 118. IRIS

California possesses a great many species
of LARKSPUR or DELPHINIUM, and many of
them are blue. A representative is *Delphinium decorum*, figure 119, with divided
leaves and spurred blue flowers. All our
native larkspurs are perennial. This species
is found on open grassy slopes near the
coast from Humboldt County to Santa
Cruz County. Other blue-flowered ones
range into the interior and south to the
border. In some parts of the country where
Larkspur is one of the first plants to send
out leaves in the early spring, it poisons a
great many cattle. See also pages 39 and 84.

FIGURE 119. BLUE LARKSPUR

Among the lupines very many are blue,
and these may be annual or perennial or
even woody. A woody species of LUPINE
becoming a yard or more tall and with long
racemes of beautiful flowers two-thirds of
an inch in length is *Lupinus longifolius*,
figure 120, found on coastal bluffs and in
canyons from Ventura County to San
Diego County. The leaves are more or less
silky.

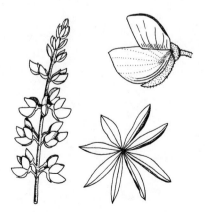

FIGURE 120. BUSH LUPINE

FIGURE 121.   SEA-LAVENDER

FIGURE 122.   GILIA TENUIFLORA

FIGURE 123.   BIRD'S EYES

In coastal salt marshes and on back beaches grows a bluish-flowered herb called SEA-LAVENDER or MARSH-ROSEMARY (*Limonium californicum*), figure 121. It has a woody base, large basal leaves, and open branched flower clusters with bluish or violet flowers about one-fourth inch long. These dry and persist, so that the plant is a kind of "everlasting." It is found the length of the state.

Under "Red Flowers" were presented several members of the Phlox Family; quite a few of this family have bluish flowers, particularly GILIA. *Gila tenuiflora*, figure 122, is bluish or purplish, with flowers about half an inch long and the leaves mostly basal. It runs through a number of forms and is found in sandy washes and canyons about Monterey Bay and in areas about the Salinas Valley.

A species with more finely dissected leaves, also annual, erect, and branched above, is BIRD'S EYES (*Gilia tricolor*), figure 123, on open grassy places and slopes of and surrounding the Central Valley of California, from Shasta County to Tulare County. Nearer the coast it occurs from Humboldt County to San Benito County. The corolla is pale to deep blue-violet, with yellow to orange in the throat and with five pairs of purple spots. See also pages 35, 87, and 89.

Among the annuals in the Phlox Family and closely related to the gilias is a group called ALLOPHYLLUM, often quite glandular and with more or less pinnately lobed leaves. Shown in figure 124 is *Allophyllum gilioides*, found on dry slopes and flats below 6,000 feet and ranging from the foothills of the Sierra Nevada to the mountains of San Diego County. The dark blue-violet flower is about one-third inch long.

Another relative of *Gilia* is NAVARRETIA, usually with spine-tipped teeth to the leaves. Among those that may have blue flowers is *N. atractyloides*, figure 125, a very glandular-hairy annual with flowers almost half an inch long. It is found mostly in dry open places in the Coast Ranges from Humboldt County to San Diego. See also page 21.

In the Waterleaf Family, which often has divided leaves and the flowers in coiled clusters, is a woody group called YERBA SANTA. In the early days it was used to make a tea for colds and asthma. These shrubs are aromatic and often have viscid leaves. The species shown in figure 126, *Eriodictyon californicum*, attains a height of five or six feet and forms large patches on dry rocky slopes and ridges. The flower is bluish-lavender to whitish and one-fourth to half an inch long; the leaves are not divided. It ranges from San Benito and Kern counties through the Coast Ranges and along the foothills of the Sierra Nevada north to Oregon. In southern California it is replaced by *E. trichocalyx* with somewhat smaller flowers.

FIGURE 124.   ALLOPHYLLUM GILIOIDES

FIGURE 125.   NAVARRETIA ATRACTYLOIDES

FIGURE 126.   YERBA SANTA

FIGURE 127. WILD-CANTERBURY-BELL

FIGURE 128. SKULLCAP

FIGURE 129. WOOLLY BLUE CURLS

In the same family as the YERBA SANTA of figure 126, the genus *Phacelia* (see pages 22 and 36) is a large and polymorphous group. Several species have somewhat bell-shaped flowers and are often called WILD-CANTERBURY-BELL, though only superficially similar to the true Canterbury-Bell (*Campanula*). Figure 127 shows one of these, *Phacelia Parryi*, a very glandular annual, one to two feet high, with coiled clusters of buds expanding into dark bluish-purple or violet flowers almost an inch across. It and related species are found in disturbed places and burns; it ranges from Monterey County to San Diego. Many people are allergic to these glandular phacelias and are affected much as by Poison-Oak.

The Mint Family, with its aromatic qualities, paired leaves, and two-lipped flowers, is represented in the blue-flowered group by a number of plants. Among these is SKULLCAP, so called because of a crest-like projection on top of the calyx. Our common one in the spring is *Scutellaria tuberosa*, figure 128, a perennial with thin underground rhizomes producing small tubers. The flowers exceed half an inch in length. The plant is quite common at the edge of brush and woods from northern to southern California, in Coast Ranges and Sierran foothills.

Another member of the family is the WOOLLY BLUE CURLS or ROMERO (*Trichostema lanatum*), figure 129, a rounded shrub to about three or four feet tall, with aromatic lance-shaped leaves and with blue flowers having long arched stamens. Most of the color, however, is in the bluish or, more often, purplish hairs of the calyx. Woolly Blue Curls ranges on dry slopes from Monterey and San Benito counties to Lower California. It has been used as an astringent and for sores and ulcers.

Sage is a member of the Mint Family and has a number of blue-flowered species. Sage, or *Salvia*, is an enormous group, ranging through California and far to the south of us. Some species, especially of the Old World, are commonly used in cookery. Among our local sages is CHIA (*Salvia Columbariae*), figure 130, a common annual of dry open disturbed places below 4,000 feet and from inner Mendocino County to Mexico. The blue flowers are about half an inch long. The seed was used extensively by the Indians for making a refreshing drink and for roasting for food.

Another much showier annual is THISTLE SAGE (*S. carduacea*), figure 131, with lavender flowers about an inch long and fringed at the edges. The stamens bear almost vermilion anthers. Thistle Sage is to be sought in sandy and gravelly places of the interior from Contra Costa, San Joaquin, and Stanislaus counties through the Central Valley and inner Coast Ranges to Kern County, then south to Lower California. It is quite spiny.

California is the home also for a number of shrubby sages; as an example is CLEVELAND SAGE, *Salvia Clevelandii*, figure 132, a fragrant shrub to about three feet tall and bearing leaves with impressed veins. The odor is very pleasant, especially in the early morning or after rain. The flowers are dark blue-violet and almost an inch long. It is a chaparral plant of San Diego County and northern Lower California. If used sparingly, its dried leaves are a good substitute for the cultivated Sage of cookery. See also pages 23 and 86.

FIGURE 130. CHIA

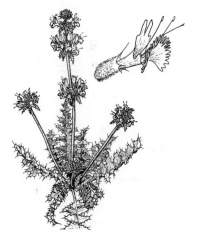

FIGURE 131. THISTLE SAGE

FIGURE 132. CLEVELAND SAGE

FIGURE 133. PURPLE NIGHTSHADE

FIGURE 134. BLUE-EYED MARY

FIGURE 135. PENSTEMON

The Nightshade Family often has open potatolike flowers. The Nightshade genus (*Solanum*) may be herbaceous or woody; an example of the latter type is S. *Xanti*, figure 133, our common PURPLE NIGHT-SHADE, which is found from San Luis Obispo County to the Mexican border. The corolla is deep violet to dark lavender, one-half to an inch in diameter, and the fruit is a small round greenish berry. The Channel Islands have a related species, S. *Wallacei*, with larger flowers and fruits (see page 79). On dry brush-covered places and in canyons from Mendocino County to Los Angeles County is another species, S. *umbelliferum*, with somewhat similar flowers but with at least some of the hairs forked and often abundant enough to give the plant an ashy appearance.

Chinese Houses (*Collinsia*) has been mentioned on another page (30); of the blue species there is illustrated here BLUE-EYED MARY (*C. Greenei*), figure 134, a slender annual of rocky and stony places below 7,000 feet and from Trinity and Humboldt counties to Sonoma and Lake counties. The flowers are almost half an inch long. *Collinsia* can always be identified by the fact that the middle lobe of the lower lip of the corolla is developed into a keel-shaped structure in which are contained the stamens. PENSTEMON has been discussed elsewhere (pages 62 and 90), but there is shown in figure 135 one of the most conspicuous of the California species, *P. spectabilis*. It is a large plant, woody below, up to four feet high, with lavender-purple flowers with blue lobes and attaining a length of an inch or more. It grows from Los Angeles County south.

In the Bellflower Family the ovary or
young seed pod is beneath the flower instead of up inside it. A member of the
family often found on burns in the chaparral and in other disturbed areas is GITHOP-
SIS SPECULARIOIDES, figure 136. It has stiff
stems to about six inches high and occurs
in both blue and white forms through the
length of the state, the white form being
the more southern.

Another member of the family is DOWN-
INGIA, but more like a lobelia and with a
two-lipped corolla, the lower lip being
quite enlarged. The ovary is long and stem-
like. *Downingia cuspidata*, figure 137, is a
species from San Luis Obispo and Madera
counties northward and only occasionally
southward. Like the other downingias it
is to be sought in drying pools after the
winter rains or in other moist places.

FIGURE 136.   GITHOPSIS

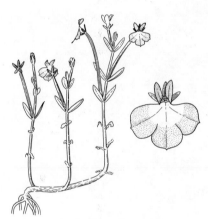

FIGURE 137.   DOWNINGIA

FUCHSIA-FLOWERED GOOSEBERRY (*Ribes speciosum*), plate 49, is a spiny shrub three to six feet tall with long spreading bristly branches, shining green leaves, and bright red hanging flowers with four sepals and four petals. In the spring it is striking plant and is often met in cultivation. It is common in shaded canyons below 1,500 feet near the coast, from Santa Clara County south. It begins to bloom shortly after the New Year. For other species of *Ribes* (Currants and Gooseberries) see pages 52 and 96.

PLATE 49. FUCHSIA-FLOWERED GOOSEBERRY

SUGAR BUSH (*Rhus ovata*), plate 50, is an evergreen shrub related to Sumac and Poison-Oak. It has stout reddish twigs, leathery leaves two to three inches long, and small reddish flower buds in dense clusters. So it is through the winter; in the spring the pinkish flowers open and are followed by acid-covered reddish fruits. It grows on dry slopes from Santa Barbara County to Lower California and is most abundant in the hot interior valleys, although it is found on Santa Cruz and Catalina islands. It has a close kin in Lemonade Berry, *Rhus integrifolia*, of coastal bluffs in southern California and can be distinguished by its ovate leaves somewhat folded along the midrib; in Lemonade Berry they are oblong and flat. Both species have an acid substance on the surface of the flattened fruits.

PLATE 50. SUGAR BUSH

One of our few Western species of TRILLIUM is *T. ovatum*, plate 51. It has three leaves in a whorl and a solitary white flower that bears three petals which turn rose in age. Found on moist shaded slopes, this Trillium is to be sought from Monterey County north, flowering from February to April. See also page 90.

PLATE 51. TRILLIUM

75

PLATE 52. MANZANITA

MANZANITA (see pages 19 and 29) has more than forty species in California. One of the most attractive is *Arctostaphylos Stanfordiana*, plate 52, of dry slopes in the chaparral from Napa and Lake counties to Mendocino County. Erect, much-branched, and three to six feet high, it is characterized by delicate pink flowers about one fourth of an inch long. Most species of course are white. There are few brushy slopes in the state without one form or another of Manzanita.

PLATE 53. MANZANITA FRUITS

The word manzanita is the Spanish for "little apple" and the fruiting cluster shown in plate 53 (*Arctostaphylos insularis*) tells why this name was applied. The fruits of some species, if not too dry, make an excellent jelly.

PLATE 54. BUSH-RUE

In the Rue Family, to which *Citrus* belongs, California has relatively little native representation. The family is characterized by its abundant oil glands in all vegetative parts and this condition is fully met in BUSH-RUE (*Cneoridium dumosum*), plate 54. A dense bush three to six feet high with slender twigs, narrow leaves and small four-petaled white flowers, it is quite an attractive plant. It grows in chaparral of Orange and San Diego counties and flowers from November to April.

On beaches and near the immediate coast from Monterey County to Lower California grows Ice Plant (*Mesembryanthemum crystallinum*), plate 55, so called because the epidermal cells are elevated and form little glistening masses that make a sparkling surface on the leaves and stems. It is a succulent annual, prostrate, with leaves to two or four inches long, and flowers almost an inch across. It blooms from March to October. Another species of *Mesembryanthemum* is on page 52.

PLATE 55. ICE PLANT

Fiddleneck with its coiled cyme of flowers and its ovary usually forming four one-seeded nutlets, is a characteristic member of the Forget-me-not Family. With its yellow to orange color it is distinctive among California members of the family. The species in the close-up pictured in plate 56 is *Amsinckia Douglasiana* of open places in the inner Coast Ranges from Monterey County to Santa Barbara and Kern counties. It is distinguishable with difficulty from other species in most of the area covered by this book. Often the plants have very stiff hairs that are quite prickly to the hand.

PLATE 56. FIDDLENECK

The field of Fiddleneck in plate 57 is *Amsinckia intermedia* var. *Eastwoodae* with much larger and more highly colored flowers showing in what profusion these plants may occur. In this species of the Great Central Valley and of much of California toward the coast, the corolla is over half an inch long and is deep orange. See also page 102.

PLATE 57. FIDDLENECK

PLATE 58. REDBUD

REDBUD (*Cercis occidentalis*), plate 58, is one of California's most beautiful shrubs in early spring. It is found on dry slopes and in canyons below 4,000 feet, in the inner Coast Ranges from Humboldt County to Solano County and in the Sierran foothills from Shasta County to Tulare County, with an outlying colony in eastern San Diego County. The flowers are almost half an inch long and nearly all of them appear before the leaves, so as to make masses of magenta to pinkish color among other shrubs. Some individuals have quite highly colored seed pods that persist through most of the summer. The Indians are said to have used the split twigs for basketry.

PLATE 59. FIRE-CRACKER FLOWER

FIRE-CRACKER FLOWER (*Brodiaea Ida-Maia*), plate 59, does not look much like most of our other brodiaeas, which are blue or yellow (see page 83). The flowers are about an inch long; the plant grows to a height of one to almost three feet. It is found in canyons and in open grassy places between 1,000 and 4,000 feet, from Lake and Shasta counties to southern Oregon, and begins to bloom in May.

PLATE 60. DESERT-DANDELION

DESERT-DANDELION (*Malacothrix californica*), plate 60, is an annual, much-branched from the base, with finely divided leaves and each branch ending in a dandelionlike head of pale yellow strap-shaped flowers. It is very fragrant. Found in dry sandy places from Sacramento Valley south, it extends its range to near the coast in Monterey and San Luis Obispo counties and occurs also on Santa Cruz and Santa Catalina islands. It blooms from March to May.

DUTCHMAN'S PIPE is a widely used name for a group of vines with a peculiar tubular calyx usually bent or curved to resemble the bowl of an old-fashioned pipe. Some of the tropical species are gorgeously colored and very large-flowered. Our California representative, *Aristolochia californica*, plate 61, is not particularly beautiful but is striking nonetheless. It is a woody climber eight to twelve feet high and usually found in low ground growing over bushes. It ranges from Monterey County north in the Coast Ranges and in the Sierran foothills from Sacramento and Eldorado counties to the head of the Sacramento Valley. It begins to flower in January. The leaves are two to six inches long.

PLATE 61. DUTCHMAN'S PIPE

PURPLE NIGHTSHADE is a close relative of the Potato and Tomato. Its flat open flowers vary from half an inch to an inch and one-half in diameter and are striking, with bright yellow anthers standing together at the center of the flower. The fruit resembles a small green tomato turning purple when ripe. The species shown in plate 62 is *Solanum Wallacei* of Santa Catalina Island. On the mainland we have S. *Xanti* (see page 72) with smaller flowers and greenish leaves and S. *umbelliferum*, which is often grayish-hairy and offers quite a contrast in color with its purple blossoms. These plants are more or less woody and found in one form or another in most of the area covered by this book.

PLATE 62. PURPLE NIGHTSHADE

SEASIDE DAISY (*Erigeron glaucus*), plate 63, is another strictly coastal plant of beaches, dunes, and sea bluffs. It is a low perennial with rather fleshy leaves and pale violet to lavender flowers. These heads are about an inch or two inches in diameter. It is found on the Santa Barbara Islands and from San Luis Obispo County to Oregon. It flowers from April to August.

PLATE 63. SEASIDE DAISY

PLATE 64.   WILD ONION

WILD ONION has almost forty species in this state. Like the cultivated onions they have a characteristic, often strong, onion-like or garliclike flavor. The one illustrated in plate 64 (*Allium praecox*), grows to a foot or so tall and bears rose-purple or lighter flowers with dark midveins. It is found in brushy or wooded places from Ventura to San Bernardino counties and on the islands off the coast. It is very similar to other species widely distributed up and down California. The bulbs were widely eaten by the Indians.

PLATE 65.   WILD ONION

In plate 65 is another species, *Allium falcifolium*, of a lower growth habit, with broad flat leaves and more numerous flowers. It grows in heavy or rocky soil, often on serpentine outcrops below 7,000 feet from Santa Cruz County to southern Oregon and flowers from March to July.

PLATE 66.   GOLD-FIELDS

SUNSHINE or GOLD-FIELDS (*Baeria chrysostoma*), plate 66, is a small annual occurring in untold numbers on open flats and slopes from San Diego County to Oregon. Often only a few inches high, it begins to bloom in March and forms great masses of color. The narrow entire leaves are opposite.

80

THISTLE (*Cirsium Coulteri*), plate 67, is a woolly spiny plant, two to four feet tall, with long, lobed leaves ending in numerous long spines. The flower head has a wonderful pattern of spirally arranged spine-tipped bracts connected with each other by a cobwebby white wool. This thistle is found on dry slopes to quite high altitudes, from Mendocino and Butte counties south, especially away from the immediate coast. Closely related species may have more or less wool, vary in stature and color, and have more local distribution. Flowers usually begin in April and continue through May or June.

PLATE 67. THISTLE

MARIPOSA-LILY is a name given to many species of *Calochortus* with erect more or less bowl-shaped flowers. The plants have underground bulbs and long linear leaves. Flowers are of many colors, often with various patterns and blotches, as can be seen in the illustration of *C. venustus*, plate 68, which grows at elevations of 1,000 to 8,000 feet on the west base of the Sierra Nevada from Eldorado to Kern counties and in the Coast Ranges from San Francisco Bay to Los Angeles County. It begins to bloom in May.

PLATE 68. MARIPOSA-LILY

Another species of Mariposa-Lily is *Calochortus clavatus*, a yellow-flowered species shown also in plate 69. It is found on dry often rocky slopes below 4,000 feet, from Eldorado to Mariposa counties in the Sierra and from Stanislaus to Los Angeles counties in the Coast Ranges. The lemon-yellow petals may have red-brown markings. The bulbs of MARIPOSA-LILIES were an important source of food for the Indians. See also page 9.

PLATE 69. MARIPOSA-LILY

PLATE 70. CALIFORNIA POPPY

PLATE 71. CALIFORNIA POPPY

PLATE 72. BLAZING STAR

CALIFORNIA POPPY (*Eschscholzia californica*), plate 70, is one of the most typical plants of the state and is, indeed, the state flower. The species can be annual to perennial and vary from a compact form among coastal dunes to a large perennial of the interior (with large deep orange flowers in early spring gradually giving way to pale yellow smaller ones in June), to an annual form in the San Joaquin Valley and southern California.

But to be *E. californica* there must be a flat rim below each flower which is quite conspicuous after the petals fall away. Other species lack this rim. Flowering begins usually in February or March. The illustration with the single flower is an inland form of *E. californica*, plate 70, the one of the entire plant is the coastal dune form (var. *maritima*), plate 71.

BLAZING STAR is the name given to the various species of Mentzelia, one of the showiest being *M. Lindleyi*, plate 72. It is an annual, one to two feet high, freely branched, with divided leaves. The flowers are one and one-half to three inches in diameter and open toward evening. It is found on sunny rocky slopes below 2,500 feet from Alameda County to Santa Clara County and western Stanislaus and Fresno counties. It blooms in April and May. Another more widely distributed species is *M. laevicaulis*, with narrower petals two to three inches long and of a paler yellow or cream color. See also page 100.

PLATE 73. ITHURIEL'S SPEAR

Another Brodiaea is GRASS NUT or ITHURIEL'S SPEAR (*Brodiaea laxa*), also shown in plate 73, and with much larger flowers in an open cluster. It is common in heavy soils below 4,600 feet, in the Coast Ranges from Oregon to San Bernardino County and in the Sierran foothills from Tehama County to Kern County. Flowering is from April to June. See also page 78.

GOLDEN BRODIAEA or PRETTY FACE (*Brodiaea lutea*), plate 74, may be confused with Golden Stars, but is mostly a paler, less golden yellow and the petallike segments are more egg-shaped. It too has basal linear leaves and a simple stem with an umbel of many flowers at the summit. It occurs from Tulare and Kern counties at the base of the Sierra Nevada and from San Luis Obispo County in the Coast Ranges north to Oregon. Flowering is from March into the summer.

PLATE 74. GOLDEN BRODIAEA

BLUE DICKS or WILD HYACINTH (*Brodiaea pulchella*), plate 75, has long naked stems with basal leaves about a foot long. The flowers are in dense heads, mostly bluish-violet. Often common in dense masses, Blue Dicks is found in open places in most parts of California west of the Sierra and the southern mountains. It blooms from March to May.

PLATE 75. WILD HYACINTH

PLATE 76. GUM PLANT

GUM PLANT (*Grindelia stricta* subspecies *venulosa*), plate 76, is a more or less
prostrate perennial that is very resinous
and sticky. It is confined to coastal salt
marshes and sea bluffs from Monterey
County north, but closely related species
are in many parts of the state. Some of
them grow more upright; all have bright
yellow flowers and are more or less viscid.
They bloom in late spring and in summer.

LARKSPUR (see pages 39 and 67) has
some thirty species in our state and many
of these have several geographical forms.
Usually the showy part of the flower is the
calyx, the posterior sepal being spurred.
The petals are small and in the center of
the flower. Most of our larkspurs are perennials and most are blue to white. The one
shown in plate 77, *Delphinium Parryi* subspecies *Blochmanae*, is a light blue with
white petals and occurs in San Luis Obispo
and Santa Barbara counties. Darker forms
of *D. Parryi* are found both to the north
and the south. Flowers appear mostly in
April and May.

PLATE 77. BLUE LARKSPUR

PLATE 78. FIVESPOT

FIVESPOT (*Nemophila maculata*), plate
78, is a low spreading annual with lobed
leaves and white flowers that open wide
and may be more than an inch across. The
five purple blotches are characteristic. It
occurs in open places along the west base
of the Sierra Nevada from Plumas County
to Kern County and flowers from April to
July. Other species of *Nemophila* may be
white or various shades of blue (Baby
Blue-Eyes, see page 89).

CALIFORNIA ROSE-BAY or RHODODENDRON (*Rhododendron macrophyllum*), plate 79, is an evergreen shrub three to ten or more feet high with coarse twigs and leathery dark green leaves. The flowers are more than an inch long, broadly bell-shaped, rose to rose-purple, rarely white. Growing in dryish to damp, more or less shaded woods below 4,000 feet, it is found near the coast from Monterey to Del Norte and Siskiyou counties and blooms from April to July. It adds much color to the woods in the Redwood region. See also pages 19 and 38.

PLATE 79. RHODODENDRON

MONKEY-FLOWER (see pages 37, 62, and 103) is represented in plate 80 by a yellow species, *Mimulus guttatus*. It is a perennial herb with more or less hollow stems, broad toothed leaves, angled calyx, and two-lipped yellow flowers usually spotted red. They are from about one-half to one and one-half inches long. The species is very variable and, in its several forms, it is found in wet places through most of California below 10,000 feet except on the actual deserts. Even there it is occasional. It flowers from March to August.

PLATE 80. MONKEY-FLOWER

TIDY TIPS is the name of *Layia platyglossa*, plate 81 (see also page 35). It is common in grassy places at low elevations from Mendocino and Butte counties south. It is conspicuous with its usually white-tipped petallike rays. The leaves are lobed. It flowers from March to June.

PLATE 81. TIDY TIPS

85

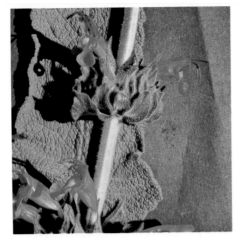

PLATE 82. PITCHER SAGE

PITCHER SAGE (*Salvia spathacea*), plate 82, is a coarse perennial herb with underground rootstocks, large deeply veined leaves, and purplish-red flowers more than an inch long. It is found in grassy and shaded places below 2,000 feet and ranges from Sonoma County to Orange County. The bracts at the base of the flowers are often quite conspicuous and mostly purplish. Flowering is from March to May.

Another notable shrub is SPICE BUSH or SWEET SHRUB (*Calycanthus occidentalis*), plate 83. It is usually rounded, three to fifteen feet tall, and pleasantly aromatic when bruised. It is deciduous, the leaves being two to five inches long. The flowers are solitary, reddish-brown, one to over two inches long. Found in moist places in canyons and about ponds, below 4,000 feet, it occurs in the North Coast Ranges from Napa County to Trinity County and along the west base of the Sierra Nevada from Tulare County to Shasta County. Flowering is from April to August. The fruit is a woody, much-veined, ovoid structure one to almost two inches long.

PLATE 83. SPICE BUSH

YARROW is an old English name for a group of aromatic herbs widely distributed in the northern hemisphere. The Yarrow shown in plate 84 is *Achillea borealis* subspecies *californica*, a perennial with finely dissected leaves and flat clusters of small white heads. It is found in open and grassy places below 2,500 feet in the Sierran foothills and in the Coast Ranges and southern California. It blooms from March to June. The leaves steeped in hot water were long considered useful for stopping blood in cuts and wounds.

PLATE 84. YARROW

GILIA (see pages 35, 68, and 89) is a
large group with many colors and with
flowers in open inflorescences or in tight
headlike clusters as demonstrated by the
blue *G. capitata* of plate 85. This variable
species is an annual with many forms and
in most of our area. It is characterized by
its finely dissected leaves and heads of blue
five-lobed flowers. It is a spring bloomer
and often occurs in large masses.

PLATE 85. GILIA

Perhaps the most common single shrub
in California is CHAMISE or GREASEWOOD
(*Adenostoma fasciculatum*), plate 86,
covering thousands of square miles of hillsides and mountain slopes below 5,000
feet. With small narrow fascicled leaves
and shreddy reddish bark, it bears in May
large panicles of minute white flowers,
each with ten to fifteen stamens. These
flowers turn brown and dry and persist
through the summer.

PLATE 86. CHAMISE

Clematis lasiantha, plate 87, is called
CLEMATIS or VIRGIN'S BOWER. It is a woody
climber ten to fifteen feet high and spreads
out over bushes in the chaparral, growing
from Trinity and Shasta counties south to
the border. The compound leaves usually
have three leaflets; the flowers are up to
about an inch in diameter and make masses
of cream color. They appear from March
to June.

PLATE 87. CLEMATIS

PLATE 88. LIVE-FOREVER

Another LIVE-FOREVER (see page 39) is *Dudleya caespitosa*, plate 88, a species found on sea bluffs from Monterey County to Los Angeles County. A plant with more powdery, broader leaves and associated with it is *D. farinosa*. The petals are about one-half of an inch long. Flowering is from April to July.

PLATE 89. HUMBOLDT LILY

TIGER LILY or HUMBOLDT LILY (*Lilium Humboldtii*), plate 89, is one of our most common and conspicuous lilies. It is a stout plant becoming several feet tall and bears whorls of ten to twenty leaves, giving a conspicuous tiered effect. The flowers are many, three to several inches across, nodding, orange-yellow with conspicuous maroon or purple spotting. This lily grows in dry rather heavy soil from Butte to Fresno counties and in a somewhat modified form from Santa Barbara County south. It begins to bloom about the first of June.

PLATE 90. GOLDEN STARS

GOLDEN STARS (*Bloomeria crocea*), plate 90, have long basal linear leaves and a simple stem with a terminal umbel of orange-yellow to pale yellow flowers. These are about an inch in diameter. The species has three forms which occur from Monterey County to Lower California. Flowering is from April to June. Being related to the lilies, the flowers have six petallike segments instead of five as in the gilias.

STREAM ORCHIS (*Epipactis gigantea*), plate 91, is an orchid with a short creeping rootstock and leafy stems. The flowers per plant commonly number three to fifteen, and are about an inch in diameter, with greenish sepals and purplish or reddish petals. The lip is strongly purple-veined. Growing in moist places below 7,500 feet, the species occurs almost throughout California. It flowers from May to August.

PLATE 91. STREAM ORCHIS

BABY-BLUE-EYES is a common name for a most variable species with a wide range of flower size and color. Usually some shade of blue, the most common species is *Nemophila Menziesii*, plate 92. It sprawls or grows up among other plants and has compound leaves with several divisions that are usually in turn toothed. The open flower can be as much as an inch or an inch and one-half across. The species is found in grassy and brushy places in much of California and blooms from February to June. See also page 84.

PLATE 92. BABY-BLUE-EYES

Collomia grandiflora, plate 93, sometimes given the name WILD BOUVARDIA, belongs to the Phlox Family, with *Gilia* and *Linanthus*. It has erect stems a foot or more high and entire lanceolate to linear leaves. The flowers are in tight terminal clusters and vary from yellow-salmon to almost white. It is found in much of California and blooms in April and May.

PLATE 93. WILD BOUVARDIA

89

PLATE 94. PENSTEMON

Penstemon is a large western genus, some species with a hairy sterile stamen being called Beard-Tongue (see pages 62 and 72). The PENSTEMON illustrated in plate 94, *P. heterophyllus*, is one of the most widespread in California. It is somewhat woody at the base, usually one to one and one-half feet high, with narrow leaves and a narrow cluster of rose-violet flowers usually with blue lobes. The gaping corolla is about one inch long. This species grows on dry hillsides below 5,000 feet from Humboldt County to San Diego County. It begins to flower in April. Its undeveloped buds are yellow, while those of a closely related blue species, *P. azureus*, are not.

PLATE 95. TRILLIUM

TRILLIUM and WAKE ROBIN are names used for many species of the genus *Trillium*, *T. chloropetalum* being shown in plate 95. The stout stem is a foot or more tall and bears a whorl of three large broad leaves and a solitary mostly greenish flower. In some forms the petals are deep red to lilac. The plant is found on brushy and wooded slopes from central California to Washington and blooms from March to May. See also page 75.

PLATE 96. ROCK-ROSE

ROCK-ROSE (*Helianthemum scoparium*), plate 96, is a bushy low perennial with many ascending to spreading stems to about one foot long. The short leaves are narrow. The flowers are yellow, about half an inch wide. In various forms, the species occurs in much of the state west of the mountains and blooms from March to early summer.

FLOWERS YELLOW TO ORANGE
OR GREENISH

Section Four

Related to False Solomon's-Seal (page 8) and with slender rootstocks and leafy stems as in that plant, but with the drooping flowers one to two instead of in elongate clusters, is Fairy Bells (*Disporum Hookeri*), figure 138. Its flowers are more or less greenish with a cream or whitish tinge, about half an inch long, and produce scarlet berries. It is found in shaded woods away from the immediate coast from Monterey County to southern Oregon.

FIGURE 138. FAIRY BELLS

In the Iris Family is a genus with grass-like leaves arranged in the two rows so typical of irises. It is *Sisyrinchium,* most species of which have blue flowers and are called Blue-eyed-Grass, but others are yellow, as the Golden-eyed-Grass (*S. californicum*), figure 139. Each segment of the yellow flowers has five to seven dark nerves. The plant is found in moist places at low elevations near the coast from Monterey County to Oregon.

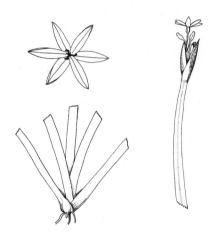

FIGURE 139. GOLDEN-EYED-GRASS

Related to our cultivated Calla-Lily is a big family, the Arum Family. It is largely tropical, but a few species have come into the colder parts of the world. All have a more or less colored and showy leaf (spathe) variously cupped or wrapped around a fleshy central stalk (spadix) that bears numerous small divisions, the individual flowers. Because many of these plants attract insects by their strong often unpleasant odor, they are commonly called Skunk-Cabbage. On the Pacific Coast we have a Yellow Skunk-Cabbage (*Lysichiton americanum*), figure 140, with a yellowish spathe and greenish spadix. These appear before the leaves, which are large and more or less elliptic in shape. The plants grow in open swamps and wet woods from the Santa Cruz Mountains to Del Norte County, and on to Alaska and Montana.

FIGURE 140. YELLOW SKUNK-CABBAGE

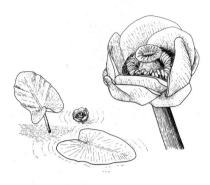

FIGURE 141.   COW-LILY

FIGURE 142.   TREE POPPY

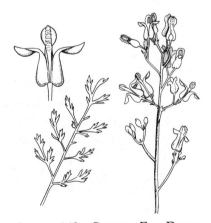

FIGURE 143.   GOLDEN EAR-DROPS

The southern parts of California have so few natural bodies of water that native water-lilies would seem most unexpected, but from San Luis Obispo and Mariposa counties northward are found two species: one of these, the COW-LILY or YELLOW POND-LILY (*Nuphar polysepalum*), figure 141, has floating leaves and fairly large yellow flowers with a diameter of about two to three inches. It is found in ponds and sluggish streams below 7,500 feet and begins to flower in April.

On pages 40 and 41 are accounts of some of the white-flowered poppies. Others have yellow flowers, notably the TREE POPPY or *Dendromecon*. This shrub is almost confined to California and runs through several forms, the outstanding two being *Dendromecon rigida*, figure 142, from the mainland and *D. Harfordii* from the Channel Islands. The former has narrow willow-like leaves, the latter broader less pointed ones. Both species are found on brushy slopes; the flowers are an inch or more in diameter and a lovely yellow.

People coming from our Eastern states are familiar with the old-fashioned pink Bleeding-Heart. In California our most striking species of its genus, *Dicentra*, is GOLDEN EAR-DROPS (*Dicentra chrysantha*), figure 143. It is a coarse perennial up to five feet tall, with large, much dissected leaves and smallish bright yellow flowers with two pairs of opposite petals, the two members of each pair facing each other as in Bleeding-Heart.

California has several species of BAR-
BERRY or MAHONIA; these are shrubs with
spiny-toothed leaves and yellow flowers in
drooping clusters or racemes. Although
small, the flowers are quite interesting with
sepals and petals and stamens arranged in
groups of three and with the anthers opening by uplifting flaps. The species shown
in figure 144 is *Berberis pinnata* that grows
on dry slopes from San Diego County
through the Coast Ranges to Oregon. The
flowers are followed by blue berries. A
closely related species is called Oregon-
Grape. Indians ate the berries and the
yellow inner bark yielded a dye.

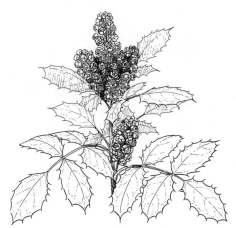

FIGURE 144. MAHONIA

We have quite a number of yellow mustards, plants with four sepals, four petals,
and a long narrow capsule or a short round
flattish one. Mustard proper or *Brassica* is
an Old-World genus that has become a
common weed in the United States. Its
fruit ends in a long point or beak. Our
early spring species, FIELD MUSTARD (*Brassica campestris*), figure 145, adds much color
to fields and orchards; its stem leaves have
clasping bases.

FIGURE 145. FIELD MUSTARD

Another group in the Mustard Family
is the WALLFLOWER; the species herewith
illustrated is found on coastal strand and
dunes in central and northern California,
Erysimum Menziesii, figure 146. The bright
yellow petals are a half-inch or longer;
other species grow taller and may be yellow or orange. All are very sweet-smelling.

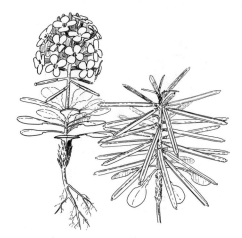

FIGURE 146. WALLFLOWER

FIGURE 147.   BLADDER POD

FIGURE 148.   GOLDEN CURRANT

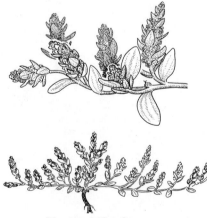

FIGURE 149.   SALTBUSH

With the same four-merous flowers as the preceding member of the Mustard Family, but with a rank unpleasant smell to the crushed leaves, is BLADDER POD (*Isomeris arborea*), figure 147. It is a low shrub with leaves divided into three leaflets, with terminal flower clusters and much inflated long-stalked pods. It is related to the Caper of cookery.

Currants and gooseberries, the genus *Ribes*, mostly have pink or red or white flowers, but a few are yellow. One, GOLDEN CURRANT (*Ribes aureum*), figure 148, is mostly east of the Sierra Nevada, but has a form in the Coast Ranges from Alameda County to Riverside County. The berry is generally orange to yellowish in the California plants.

Along the sandy beaches are several species of SALTBUSH, which tend to be prostrate annuals or perennials. The one figured, *Atriplex leucophylla*, figure 149, is white-scurfy, with leaves an inch or more long. The seed is borne between two spongy bracts that usually have warty projections on their surfaces. Other species have these bracts reddish and fleshy; still others, usually of salty or alkaline places in the interior, may be sizable roundish shrubs. But always they are scurfy and bear their seed between pairs of bracts.

Potentillas have strawberrylike flowers that may be yellow or white or even red. They belong to the Rose Family. One species found along coastal strand and about salt marshes from Los Angeles County to Alaska is PACIFIC SILVERWEED (*Potentilla Egedei*), figure 150. It has pinnately compound leaves (leaflets borne along the sides of a central axis) and runners that produce yellow flowers and start new plants. The flowers are about an inch in diameter.

Another is STICKY POTENTILLA (*P. glandulosa*), figure 151, a bushy perennial, also with pinnate leaves and quite sticky because of the glands it bears. The flowers are pale yellow to cream. It is common in brushy places and canyons the entire length of California.

Some members of the Pea Family are yellow-flowered, especially the genus *Lotus*, of which we have many species, some prostrate, some bushy and erect. A common name often given is BIRD'S FOOT TREFOIL; the species figured is *Lotus grandiflorus*, figure 152. It is one to two feet high, with seven to nine leaflets arranged pinnately and with tight clusters of yellow flowers aging red and almost an inch long. It grows on dry slopes below 6,000 feet, mostly in brushy places and the length of the state.

FIGURE 150.   PACIFIC SILVERWEED

FIGURE 151.   STICKY POTENTILLA

FIGURE 152.   BIRD'S FOOT TREFOIL

FIGURE 153.   FALSE-LUPINE

FIGURE 154.   YELLOW VIOLET

FIGURE 155.   JOHNNY-JUMP-UP

Another member of the foregoing Pea Family with yellow flowers is FALSE-LUPINE, stout perennial herbs with compound leaves having three leaflets arising at a common point (palmate). California has three species, the one shown here is *Thermopsis macrophylla*, figure 153. It grows in the higher hills in eastern San Diego County and at somewhat lower altitudes farther north and into Oregon. The bright yellow flowers are about two-thirds of an inch long.

The violet is known to every flower lover and has such characteristic flowers that it cannot be mistaken for anything else. California has about twenty-three species that vary widely in color and in leaves. One of the YELLOW VIOLETS is *Viola Sheltonii*, figure 154, with leaves divided. It grows in brushy or more or less wooded places at 2,500 to 8,000 feet, from Orange County north to Washington. It begins flowering in April. The two upper petals are brownish-purple streaks.

Another violet is WILD PANSY or JOHNNY-JUMP-UP (*Viola pedunculata*), figure 155, common on grassy slopes below 2,500 feet, from Sonoma and Colusa counties to Lower California. Its petals too have some brown on the back and brownish streaks on the face of the lower.

FIGURE 156. BERMUDA-BUTTERCUP

Wood-Sorrel or Oxalis is a large group of plants widely scattered over the surface of the earth. Usually each leaf has three leaflets or divisions and the sap is acid. Some species produce very handsome flowers and are prized in the garden. One, sometimes called BERMUDA-BUTTERCUP (*Oxalis Pes-Caprae*), figure 156, is an introduced weed and is becoming very common in orchards and fields, where its bright yellow flowers add much color to the winter landscape. In the summer it disappears and exists as a deep rootstock with scaly bulbs, by which underground means it spreads rapidly. It is native to South Africa. The flowers are about an inch across; other yellow-flowered native species are not nearly so showy.

Those of us interested in wildflowers cannot help occasionally wanting to know the name of given shrubs or trees, and while most of these are not contained in this small book, a few should be. Such a one is HOP TREE (*Ptelea crenulata*), figure 157, with leaves consisting of three leaflets, with small greenish-white flowers and with flat, winged fruits. It belongs to the same family as Citrus and has the glands characteristic of the family, but in this instance the odor is rather strong. Growing to a height of six to sixteen feet, Hop Tree is found in canyons and flats below 2,000 feet, from Santa Clara and Tulare counties to Shasta County.

FIGURE 157. HOP TREE

Another woody plant of importance is BIG-LEAF MAPLE or CANYON MAPLE (*Acer macrophyllum*), figure 158, a tree up to sixty feet tall, with large deeply lobed leaves. Its small greenish flowers develop later in the season into large, winged fruits. It is found throughout our area at elevations below 5,000 feet, except in valley plains.

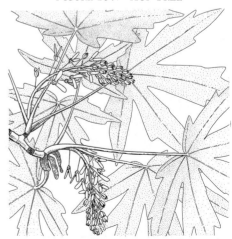

FIGURE 158. CANYON MAPLE

FIGURE 159.   BLAZING STAR

FIGURE 160.   SUN-CUP

FIGURE 161.   SILK-TASSEL BUSH

BLAZING STAR is a name given to a rather large group of plants whose surface has hairs barbed along the sides so that pieces of the plant adhere easily to cloth and to suede jackets in a most annoying manner. Some species are light yellow, some deep yellow to orange. Some have minute flowers, some quite large. Some are annual, some perennial. The one shown in figure 159, *Menzelia pectinata*, has the characteristic lobed leaves and five-petaled flowers. Blazing Stars are found throughout California except at the higher altitudes.

Resembling the Blazing Stars in having the seed pod at the base of the flower instead of up inside it, are the Evening-Primrose and SUN-CUP. They have four petals instead of five. Some are tall weedy plants, others small and almost buttercup-like. Such a one is *Oenothera ovata*, figure 160, with flowers to about an inch across. It is found in open grassy places from San Luis Obispo County to Oregon.

SILK-TASSEL BUSH is a name applied to a number of California shrubs with small greenish flowers in long hanging tassels. They are found widely distributed in brushy places; the one shown, *Garrya Congdoni*, figure 161, occurs from San Benito and Mariposa counties north.

We are all familiar with Parsnip, Celery, Carrot, and Dill, and if we have seen these plants growing, we know that their flowers are small and arranged in large compound clusters. Furthermore they have oil tubes with characteristic essential oils and odors and flavors. This Carrot Family is large in California and two spring bloomers with yellow flowers are TAUSCHIA ARGUTA, figure 162, and LOMATIUM UTRICULATUM, or WILD PARSLEY, figure 163. I know of no good common names for them, but people tend to call them Wild Parsnip or something similar. The former species is a common perennial from Santa Barbara County south.

This species in figure 163, described in the preceding paragraph, may be found from British Columbia to Lower California.

When we say the word Gentian, we think of blue, but there is a GREEN GENTIAN, mostly coarser, taller, and not so beautiful. In detail, however, the flower is interesting, having conspicuous fringed glands on the petals. The one shown here is *Frasera Parryi*, figure 164, named for Dr. C. C. Parry, who collected specimens in southern California in 1876. The species is southern, found in dry places from Los Angeles and San Bernardino counties to Lower California. It may grow to be about four feet tall and is found on open brushy slopes. Other species occur farther north.

FIGURE 162. TAUSCHIA

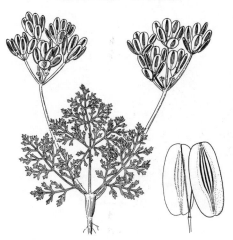

FIGURE 163. LOMATIUM

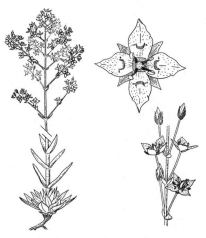

FIGURE 164. GREEN GENTIAN

FIGURE 165.   PIMPERNEL

FIGURE 166.   FIDDLENECK

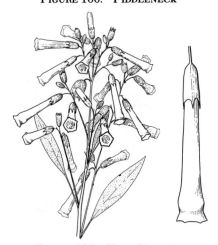

FIGURE 167.   TREE TOBACCO

One of our most common garden weeds and one that is a persistent little pest, introduced from Europe, is PIMPERNEL (*Anagallis arvensis*), figure 165. It is annual, almost prostrate, with opposite (paired) leaves and mostly salmon flowers about the size of a dime. Sometimes the flowers are bluish, sometimes almost white.

Under "White flowers" there were presented a number of plants (pages 22 and 23) with the tips curling under and forming peculiar coiled clusters of small flowers. Related to these is the yellow or orange FIDDLENECK (*Amsinckia spectabilis*), figure 166, one to two feet tall, prickly-haired, and found along the immediate coast from San Diego County to Oregon. See also page 77 for color. Other species are found in fields and grassy places in the interior.

Tobacco is largely American; one of the most conspicuous species now found in California, having arrived here from Argentina probably via Mexico, is TREE TOBACCO (*Nicotiana glauca*), figure 167, a tall shrub or small tree with pendulous tubular flowers an inch or more long. It is rather weedy, yet somewhat attractive, and it is certainly liked by humming birds. Our native tobaccos mostly have white spreading flowers and were used by the Indians for ceremonial purposes, but they are so glandular and viscid that they are not very attractive.

In the same family with the Beard-Tongue and Snapdragon is the MONKEY-FLOWER, of which California has nearly eighty species. The common yellow one of moist places in most of the state below 10,000 feet is *Mimulus guttatus*, figure 168. It has many forms with different flower sizes, is often spotted red, and may grow to a height of two to three feet. See also pages 37, 62, and 85. It is rather an attractive garden plant. As in other monkey-flowers the calyx is strongly angled and the corolla is two-lipped.

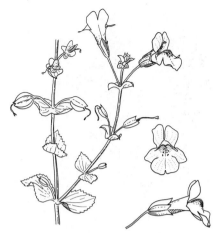

FIGURE 168. MONKEY-FLOWER

In this same family are Owl's-Clover and relatives (see page 31), with small flowers in clusters near the tips of slender stems. Some species are yellow; one called JOHNNY-TUCK or BUTTER-AND-EGGS (*Orthocarpus erianthus*), figure 169, is an annual up to about a foot high and with flowers almost half an inch long. It often grows in great masses, in open grassy places particularly in and about the great Central Valley of California, but extends well beyond that into the Coast Ranges.

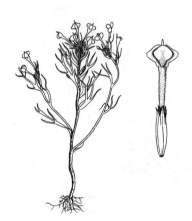

FIGURE 169. JOHNNY-TUCK

Among the honeysuckles is one found in somewhat different forms in the Sierra Nevada and along the coast. It is the TWIN-BERRY (*Lonicera involucrata*), figure 170, with paired yellowish tubular flowers followed by rather persistent black berries surrounded by conspicuous bracts. It grows to be about five to ten feet high and is found from Santa Barbara and Tulare counties north almost to the Arctic.

FIGURE 170. TWINBERRY

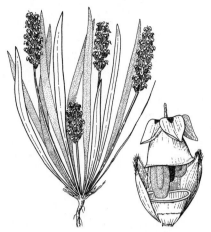

FIGURE 171. PLANTAIN

FIGURE 172. GROUNDSEL

FIGURE 173. GROUNDSEL

A common lawn weed in most of the United States is Plantain, with either narrow or broad leaves in a basal rosette and the flowers in dense spikes at the ends of leafless erect stems. These weeds were introduced from Europe. California and the Southwest have a series of small annual species, usually soft-hairy and rather dainty plants with small greenish flowers. They occur in great numbers at low altitudes, especially in sterile soil, and are found in most of the state. The PLANTAIN illustrated in figure 171 is a form of *Plantago Hookeriana*. Note how the top of the seed capsule lifts off like a lid.

In the Sunflower Family what appears to be a single flower is in reality a whole cluster of small flowers packed tightly together in a head. The outer ones may be petallike, the inner are usually tubular. Generally such flowers are called daisies, but there is an infinite variety of kinds. Among these come the GROUNDSEL or RAGWORT of the genus *Senecio*. In figure 172 is shown *Senecio aronicoides*, a perennial herb one to three feet tall and found from San Mateo County north. Other species farther south much resemble this one. A common name sometimes used is Old Man because of the white tuft of hair on each seed. SENECIO CALIFORNICUS, figure 173, is an annual species of the previously described Sunflower Family one to one and one-half feet high, with pretty yellow heads about an inch across. It occurs in dry open places from Monterey and Tulare counties south.

MAPS

TOPOGRAPHIC MAP OF CALIFORNIA
The area included in the Spring Wildflower book is that
west of the Sierra Nevada and Mojave and Colorado deserts.

COUNTY MAP OF CALIFORNIA

INDEX TO COLOR PLATES

(References are to plate numbers)

Flowers Bluish

Flowers Yellowish to Orange

INDEX

(References are to page numbers)